Geotechnical Engineering in Construction Principles and Practices for Engineers

SAKI

Geotechnical Engineering in Construction Principles and Practices for Engineers

Copyright © 2023 by SAKI

The first edition was published in 2023

ISBN:
Published by:
Sunshine
1663 Liberty Drive
Hyderabad, IN 47403
www.Sunshinepublishers.com

This book is self-published using on-demand printing and publishing, which allows it to be printed and distributed globally

Table OF CONTENT

Chapter 1: Introduction to Geotechnical Engineering 10

Overview of Geotechnical Engineering

Importance of Geotechnical Engineering in Construction Projects

Role of Geotechnical Engineers

Historical Developments in Geotechnical Engineering

Challenges and Opportunities in Geotechnical Engineering

Chapter 2: Soil Mechanics Fundamentals 20

Soil Composition and Classification

Physical Properties of Soils

Soil Water Interaction

Soil Compaction and Consolidation

Shear Strength of Soils

Soil Stress Analysis

Chapter 5: Retaining Structures 57

Chapter 6: Slope Stability Analysis 67

Chapter 7: Earthquake Engineering and Geotechnical Considerations 77

Chapter 8: Ground Improvement Techniques 87

Chapter 9: Geotechnical Aspects of Construction Materials 99

Chapter 10: Geotechnical Risk Assessment and Management 109

Chapter 11: Sustainability in Geotechnical Engineering 119

Chapter 12: Case Studies in Geotechnical Engineering 130

Chapter 13: Emerging Technologies in Geotechnical Engineering 140

Chapter 14: Professional Ethics and Responsibilities in Geotechnical Engineering 150

Importance of Professional Ethics in Engineering

Code of Ethics for Geotechnical Engineers

Professional Responsibilities in Geotechnical Engineering

Ethical Decision Making in Geotechnical Engineering

Case Studies on Ethical Issues in Geotechnical Engineering

Chapter 15: Future Perspectives in Geotechnical Engineering 160

Current Challenges and Future Trends in Geotechnical Engineering

Advances in Geotechnical Engineering Research

Technological Innovations in Geotechnical Engineering

Education

Chapter 1: Introduction to Geotechnical Engineering

Overview of Geotechnical Engineering

Geotechnical engineering is a crucial discipline in the field of civil engineering that deals with the behavior and properties of soil and rock materials. It plays a significant role in the construction industry, providing engineers with essential knowledge and tools to design and construct safe and reliable infrastructure.

This subchapter aims to provide engineers, particularly those specializing in civil engineering, with a comprehensive overview of geotechnical engineering. It will cover the fundamental principles, practices, and techniques employed in this field, highlighting their importance in construction projects.

To begin with, the subchapter will delve into the basics of geotechnical engineering, explaining the significance of understanding soil and rock mechanics. Engineers will gain insights into the composition of soil and rock, their physical and mechanical properties, and how these properties influence the behavior of the ground.

The subchapter will then explore the various geotechnical investigations and testing methods employed to gather crucial data about the site conditions. Engineers will learn about the different in-situ and laboratory tests used to determine soil and rock characteristics, including density, strength, permeability, and compressibility. Understanding these properties is crucial for making informed decisions regarding foundation design, slope stability, and overall project feasibility.

Furthermore, the subchapter will discuss the design principles and practices employed in geotechnical engineering. Engineers will be introduced to various design methods for foundations, retaining structures, embankments, and slopes. They will also learn about the different types of geotechnical engineering analyses, such as settlement analysis, bearing capacity analysis, and slope stability analysis.

The importance of geotechnical considerations in construction projects will also be emphasized. Engineers will understand how geotechnical engineering plays a vital role in ensuring the safety and stability of structures, as well as mitigating potential risks. Additionally, geotechnical challenges associated with different types of soil and rock formations will be discussed, providing engineers with valuable insights for addressing specific site conditions.

In conclusion, this subchapter provides engineers specializing in civil engineering with a comprehensive overview of geotechnical engineering. By understanding the fundamentals, investigation methods, design principles, and construction considerations in this field, engineers will be equipped to undertake geotechnical engineering projects with confidence and ensure the successful and safe completion of construction projects.

Importance of Geotechnical Engineering in Construction Projects

Geotechnical engineering plays a crucial role in the successful execution of construction projects. It involves the study of the behavior of soil and rocks that support the built environment. This subchapter aims to highlight the significance of geotechnical engineering in construction projects, specifically tailored for engineers in the niche of civil engineering.

One of the primary reasons geotechnical engineering is essential is its ability to provide valuable information about the ground conditions at a construction site. The properties of soil and rocks, such as strength, permeability, and compressibility, have a direct impact on the design and stability of structures. By conducting thorough geotechnical investigations, engineers can gather data to make informed decisions about foundation design, excavation, and slope stability. This knowledge ensures that structures are built to withstand the forces imposed on them over time.

Furthermore, geotechnical engineering helps engineers mitigate potential risks and hazards associated with the ground conditions. Through careful assessment of soil and rock properties, geotechnical engineers can identify potential issues such as landslides, liquefaction, or sinkholes. By understanding these risks, engineers can implement appropriate design measures and construction techniques to minimize or eliminate potential dangers. This not only ensures the safety of the structures but also safeguards the lives of the workers and the public.

Geotechnical engineering also plays a vital role in optimizing the cost and time efficiency of construction projects. Through detailed site investigations, engineers can identify the most suitable foundation systems and construction methods for a particular project. This knowledge allows them to make informed decisions about the most cost-effective solutions, taking into account the project's specific requirements and constraints. By optimizing the design and construction processes, geotechnical engineering helps in reducing unnecessary delays and cost overruns.

In conclusion, geotechnical engineering forms the backbone of any successful construction project in the field of civil engineering. By studying and understanding the ground conditions, engineers can design safe and stable structures, mitigate potential risks, and optimize cost and time efficiency. It is imperative for engineers to recognize the importance of geotechnical engineering and collaborate closely with geotechnical specialists throughout the entire lifecycle of a construction project. This synergy between civil and geotechnical engineers will ultimately lead to the successful completion of projects that meet the highest standards of safety, durability, and functionality.

Role of Geotechnical Engineers

Geotechnical engineers play a critical role in the field of civil engineering and construction. Their expertise lies in understanding the behavior of soil, rock, and other earth materials, and using this knowledge to design structures that can withstand the forces and conditions present at a construction site. This subchapter will delve into the important role geotechnical engineers play in the construction industry and the principles and practices they employ to ensure the safety and stability of structures.

The primary responsibility of geotechnical engineers is to assess the subsurface conditions at a construction site. They conduct thorough investigations, including soil and rock testing, to determine the physical and mechanical properties of the materials present. By understanding these properties, geotechnical engineers can predict how the ground will behave under various loadings, such as the weight of a building or the force of an earthquake.

Based on their findings, geotechnical engineers work closely with other professionals, such as structural engineers and architects, to design foundations that adequately support the proposed structure. They calculate the bearing capacity of the soil, evaluate potential settlement issues, and design appropriate measures to mitigate any geotechnical hazards. These may include techniques like soil stabilization, ground improvement, or the use of specialized foundation systems.

During construction, geotechnical engineers provide ongoing support and supervision to ensure that the design is implemented correctly. They monitor the construction process, perform quality control tests, and make any necessary adjustments to the design to accommodate unforeseen circumstances.

Geotechnical engineers also play a crucial role in assessing the environmental impact of construction projects. They evaluate the potential risks of soil erosion, slope stability, or contamination of groundwater and develop strategies to minimize these risks. They may also be involved in the design of retaining walls, embankments, or drainage systems to protect structures from natural hazards, such as landslides or floods.

In summary, geotechnical engineers are vital in the field of civil engineering and construction. They use their expertise in soil mechanics and engineering geology to ensure the safety and stability of structures. By analyzing the subsurface conditions, designing appropriate foundations, and monitoring construction processes, geotechnical engineers contribute to the successful completion of projects while minimizing potential risks. Their work is essential for creating safe and sustainable infrastructure that meets the needs of society.

Historical Developments in Geotechnical Engineering

Geotechnical engineering is an essential discipline within the field of civil engineering that deals with the behavior of earth materials and their interactions with structures and infrastructure. Over the centuries, this field has witnessed several significant historical developments that have shaped its principles and practices. Understanding the historical context of geotechnical engineering can provide engineers with valuable insights into the evolution of this discipline and the advancements it has undergone.

One of the earliest historical developments in geotechnical engineering can be traced back to ancient civilizations. The construction of large structures such as pyramids and aqueducts required an understanding of soil mechanics and foundation design. Ancient engineers developed techniques to stabilize slopes, construct foundations, and manage groundwater, laying the foundation for modern geotechnical engineering.

The Industrial Revolution in the 18th century brought about a new era of engineering advancements. Engineers like Karl von Terzaghi and Arthur Casagrande made significant contributions to soil mechanics and geotechnical engineering. Terzaghi's work on consolidation theory and the concept of effective stress revolutionized soil mechanics, providing engineers with a quantitative understanding of soil behavior. Casagrande's development of the soil classification system enabled engineers to categorize soils based on their characteristics and engineering behavior.

The 20th century witnessed further advancements in geotechnical engineering, particularly in the areas of soil testing and foundation design. The introduction of laboratory testing equipment and techniques allowed engineers to better understand soil properties and behavior under different loading conditions. This led to the development of innovative foundation systems such as deep foundations, pile foundations, and soil improvement methods.

The advent of computer technology in the late 20th century revolutionized geotechnical engineering practices. Numerical modeling and simulation tools enabled engineers to analyze complex geotechnical problems and predict the behavior of soil-structure systems. This led to more accurate and efficient design solutions, ensuring the safety and stability of infrastructure projects.

Today, geotechnical engineering continues to evolve with the integration of advanced technologies such as remote sensing, geospatial analysis, and artificial intelligence. These developments have enhanced the capabilities of engineers in assessing and managing geotechnical risks, optimizing foundation designs, and mitigating the impact of natural hazards.

Understanding the historical developments in geotechnical engineering is crucial for civil engineers. It provides a foundation of knowledge and experience that can guide engineers in making informed decisions and solving complex geotechnical problems. By studying the evolution of this discipline, engineers can gain valuable insights into the principles and practices that have shaped geotechnical engineering into what it is today.

Challenges and Opportunities in Geotechnical Engineering

Geotechnical engineering plays a crucial role in the construction industry, particularly in the field of civil engineering. This subchapter aims to shed light on the challenges and opportunities that engineers face in this specialized branch of engineering.

One of the primary challenges in geotechnical engineering is dealing with the inherent variability of soil and rock properties. The ground conditions at construction sites can vary significantly, posing challenges in terms of stability, settlement, and the overall performance of structures. Engineers must carefully analyze and interpret site investigation data to accurately assess the ground conditions and design appropriate foundations and retaining structures.

Another challenge is the increasing complexity of construction projects. With the continuous growth of urban areas, engineers are often faced with the task of constructing structures on challenging sites, such as steep slopes or reclaimed land. These projects require innovative engineering solutions and a deep understanding of soil mechanics to ensure the safety and durability of the structures.

Furthermore, the rapid advancement of technology presents both challenges and opportunities in geotechnical engineering. On one hand, engineers must keep up with the latest software and equipment to improve their analysis and design capabilities. On the other hand, these advancements open up opportunities for more accurate site investigations, advanced monitoring systems, and the use of geosynthetics to enhance soil stabilization and improve construction efficiency.

Climate change also poses significant challenges to geotechnical engineering. Rising sea levels, increased rainfall, and extreme weather events can have a significant impact on the stability of slopes, embankments, and coastal structures. Engineers must take into account these climate-related factors and incorporate them into their design and construction strategies to ensure the long-term resilience of infrastructure.

Despite these challenges, geotechnical engineering also presents numerous opportunities for engineers. The need for sustainable construction practices has led to the development of geotechnical solutions that minimize environmental impact, such as the use of geothermal energy and the implementation of green infrastructure. Additionally, the increasing demand for infrastructure in developing countries opens up opportunities for engineers to contribute to projects that improve the quality of life for communities around the world.

In conclusion, geotechnical engineering in the field of civil engineering presents a range of challenges and opportunities. By embracing technological advancements, adopting sustainable practices, and understanding the complexities of soil mechanics, engineers can overcome these challenges and contribute to the development of safe, resilient, and sustainable infrastructure.

Chapter 2: Soil Mechanics Fundamentals

Soil Composition and Classification

In the field of geotechnical engineering, understanding the composition and classification of soil is of paramount importance. As civil engineers, it is crucial to have a comprehensive knowledge of the various types of soil and their properties in order to make informed decisions during construction projects. This subchapter delves into the fundamental concepts of soil composition and classification, providing engineers with the necessary tools to assess and manage soil conditions effectively.

The composition of soil refers to the different constituents that make up its structure. These constituents typically include solid particles, water, and air. The solid particles can be classified into three primary categories: gravel, sand, and clay. Gravel consists of coarse particles, whereas sand particles are smaller in size, and clay particles are the finest. The proportions of these constituents determine the soil's overall composition and affect its engineering properties.

Soil classification, on the other hand, involves categorizing soils based on their physical and mechanical properties. The Unified Soil Classification System (USCS) is widely used in civil engineering to classify soils into distinct groups. This system considers factors such as grain size distribution, plasticity, and compressibility to assign soils into various groups, including sands, silts, clays, and organic soils.

Understanding soil composition and classification is vital in geotechnical engineering as it directly influences the behavior of soils under different loading and environmental conditions. For instance, the presence of a high clay content in a soil can significantly impact its strength and compressibility, making it more prone to settlement and instability. By assessing the soil composition and classification, engineers can make informed decisions regarding site selection, foundation design, and construction methods.

Moreover, the knowledge of soil composition and classification allows engineers to accurately predict soil behavior, such as its permeability, shear strength, and settlement characteristics. This information is crucial in designing earthworks, retaining structures, and foundations that can withstand the anticipated loads and environmental conditions.

In conclusion, the subchapter on soil composition and classification provides civil engineers with essential knowledge to analyze and interpret soil properties effectively. By understanding the composition and classification of soil, engineers can make informed decisions during the design and construction phases of a project, ensuring the safety and stability of structures. It is imperative for engineers to continually update their knowledge on soil composition and classification to adapt to the ever-evolving challenges in the field of geotechnical engineering.

Physical Properties of Soils

In the field of geotechnical engineering, understanding the physical properties of soils is of utmost importance for civil engineers. Soils, being the foundation on which structures are built, require a comprehensive analysis of their physical characteristics to ensure the safety and stability of construction projects. This subchapter aims to provide engineers in the field of civil engineering with an in-depth knowledge of the physical properties of soils and their significance in construction practices.

One of the fundamental physical properties of soils is particle size distribution. Engineers need to be familiar with the different soil particle sizes, ranging from clay, silt, sand, and gravel, as they greatly influence the behavior and strength of soils. This information is crucial when selecting suitable soil types for various construction purposes, as it determines factors such as drainage, compaction, and permeability.

Another important physical property is soil compaction. Understanding the compaction characteristics of soils is vital for engineers in order to prevent settlement and deformation issues in construction projects. This subchapter will delve into the principles of soil compaction, including factors affecting compaction, types of compaction equipment, and appropriate compaction techniques.

Porosity and permeability are also significant physical properties that engineers must consider. Porosity refers to the void space within soil particles, while permeability indicates the ease with which water can flow through the soil. Both properties are essential in designing proper drainage systems and preventing water-related damage to structures.

Additionally, soil moisture content plays a crucial role in construction practices. Engineers need to understand the relationship between moisture content and soil behavior, as it affects the soil's strength, volume change, and susceptibility to erosion. This subchapter will provide engineers with the necessary knowledge to accurately determine and control soil moisture content during construction.

Lastly, soil density is a critical physical property that engineers must assess. By measuring the density of soils, engineers can evaluate their strength and stability, as well as predict their behavior under different loading conditions. This subchapter will explore various methods for measuring soil density and their applications in construction projects.

In conclusion, the physical properties of soils are essential for civil engineers to ensure the success and safety of construction projects. This subchapter will equip engineers with a comprehensive understanding of soil particle size distribution, compaction, porosity, permeability, moisture content, and density. By mastering these physical properties, engineers will be able to make informed decisions and implement appropriate techniques in their geotechnical engineering practices.

Soil Water Interaction

In the field of geotechnical engineering, understanding the interaction between soil and water is crucial for successful construction projects. Soil water interaction plays a significant role in determining the stability, strength, and behavior of soil, making it a fundamental aspect of civil engineering.

Water is a powerful force that can drastically impact the properties of soil. It can cause soil erosion, affect the stability of slopes, induce settlements, and even lead to the failure of structures. It is, therefore, essential for engineers to comprehend the intricate relationship between soil and water to design and construct long-lasting and reliable infrastructure.

One of the key aspects of soil water interaction is the water content in soil. The water content directly influences the soil's strength and volume. Engineers must accurately measure and control the water content to ensure optimal soil conditions. Excessive water content can lead to soil saturation, reducing its strength and causing instability. On the other hand, insufficient water content can result in soil shrinkage, compromising the stability of foundations and structures.

Another crucial factor in soil water interaction is the permeability of soil. Permeability refers to the ability of soil to allow water to flow through it. The permeability of soil depends on various factors such as particle size, void ratio, and soil structure. Engineers must assess the permeability of soil to understand how water will flow through it and mitigate potential issues such as soil liquefaction or seepage.

Understanding soil water interaction also helps engineers in managing groundwater. Groundwater levels can significantly impact construction projects, particularly those involving excavations or foundations. Engineers need to assess the groundwater table and its influence on the stability of the soil and the structures to be built.

To accurately analyze soil water interaction, engineers utilize various laboratory and field tests. These tests help determine the soil's hydraulic conductivity, water retention characteristics, and its response to changes in water content. Additionally, advanced numerical modeling techniques are employed to simulate the behavior of soil under different water conditions.

In conclusion, soil water interaction is a vital aspect of geotechnical engineering, particularly in the field of civil engineering. Understanding how water interacts with soil is essential for designing and constructing safe and durable infrastructure. By accurately measuring and controlling water content, assessing soil permeability, and managing groundwater, engineers can ensure the stability and reliability of construction projects.

Soil Compaction and Consolidation

In the field of geotechnical engineering, soil compaction and consolidation are critical processes that engineers must understand and apply in construction projects. These processes play a vital role in ensuring the stability and durability of civil engineering structures.

Soil compaction refers to the process of increasing the density of soil by removing air voids and reducing its volume. It involves the application of mechanical energy to the soil mass through various compaction techniques, such as rolling, vibrating, and tamping. Compaction is essential for improving the load-bearing capacity of soil, reducing settlement, and preventing soil erosion. By increasing soil density, engineers can enhance the overall stability and performance of foundations, embankments, roadways, and other structures.

Consolidation, on the other hand, refers to the gradual settlement of soil under sustained loading. It occurs due to the expulsion of water from void spaces within the soil mass, leading to a reduction in volume. Consolidation is a time-dependent process that can significantly impact the long-term performance of structures. Engineers must consider consolidation effects during the design and construction phases to ensure the stability and serviceability of foundations and other geotechnical elements.

Understanding the principles and practices of soil compaction and consolidation is crucial for engineers involved in civil engineering projects. By effectively compacting soil, engineers can achieve the desired engineering properties, such as increased bearing capacity and reduced compressibility. They must also consider factors like soil type, moisture content, compaction energy, and compaction equipment selection to optimize the compaction process.

Similarly, engineers need to assess the consolidation characteristics of soil to accurately predict settlement and deformation behavior. They rely on laboratory testing and field monitoring to determine consolidation parameters, such as coefficient of consolidation and pre-consolidation pressure. With this knowledge, engineers can design appropriate foundation systems that accommodate anticipated settlements and minimize the risk of structural damage.

Moreover, engineers must be familiar with the various geotechnical testing methods used to evaluate soil compaction and consolidation. These tests include the Proctor compaction test, standard and modified compaction tests, and oedometer tests. Proper interpretation of test results allows engineers to make informed decisions regarding soil improvement techniques, such as dynamic compaction, vibro-compaction, and preloading.

In conclusion, soil compaction and consolidation are fundamental processes in geotechnical engineering that directly impact the stability and performance of civil engineering structures. Engineers must possess a comprehensive understanding of these processes to ensure the safe and cost-effective construction of foundations, embankments, roadways, and other geotechnical elements. By applying appropriate compaction and consolidation techniques and considering site-specific conditions, engineers can optimize the engineering properties of soil and enhance the overall performance and longevity of structures.

Shear Strength of Soils

In the field of geotechnical engineering, understanding the shear strength of soils is of utmost importance for civil engineers. Shear strength refers to the soil's ability to resist deformation and maintain its integrity under the action of external forces. It plays a crucial role in the design and construction of various geotechnical structures such as retaining walls, foundations, slopes, and embankments.

The shear strength of soils is influenced by several factors, including soil type, grain size distribution, moisture content, and the presence of any cohesive materials. It is typically described by two parameters: cohesion (C) and angle of internal friction (φ). Cohesion represents the shear strength of cohesive soils and is a measure of the inter-particle forces that hold the soil particles together. On the other hand, the angle of internal friction characterizes the shear strength of cohesionless soils and is a measure of the soil's resistance to shearing due to friction between particles.

Determining the shear strength of soils is crucial for assessing the stability of geotechnical structures. Engineers employ various laboratory and field tests to measure the soil's shear strength. One commonly used laboratory test is the direct shear test, where a soil sample is subjected to a shearing force. By measuring the force required to cause failure, engineers can determine the cohesion and angle of internal friction.

Another widely used test is the triaxial test, which simulates the stress conditions in the field more accurately. In this test, a soil sample is confined in a cylindrical chamber and subjected to axial and radial stresses. By varying the stress conditions, engineers can determine the shear strength parameters for different stress levels.

Understanding the shear strength of soils is crucial for designing safe and stable geotechnical structures. Engineers must consider the shear strength parameters when analyzing slope stability, designing foundations, and determining the bearing capacity of soils. By accurately assessing the shear strength, engineers can ensure that the structures they design can withstand the forces imposed by the soil and remain stable over time.

In conclusion, the shear strength of soils is a fundamental concept in geotechnical engineering, particularly for civil engineers. It governs the stability and safety of various geotechnical structures. By understanding the factors influencing shear strength and employing appropriate testing methods, engineers can design structures that can withstand the shear forces imposed by the soil.

Soil Stress Analysis

In the field of geotechnical engineering, understanding the behavior of soil under different stress conditions is crucial in ensuring the stability and safety of structures. Soil stress analysis is a fundamental concept that engineers, especially civil engineers, need to grasp to effectively design and construct various types of infrastructure.

Soil stress refers to the forces exerted on soil particles due to external loads. These loads can be applied by natural phenomena, such as the weight of overlying soil, water pressure, or by man-made structures like buildings, bridges, or retaining walls. Soil stress analysis involves evaluating the distribution and magnitude of these forces within the soil mass.

One of the key parameters in soil stress analysis is the effective stress. Effective stress represents the interparticle forces that hold soil particles together. It is the difference between the total stress, which includes the weight of the soil and any applied external loads, and the pore water pressure within the soil. Understanding effective stress is critical as it governs the shear strength and deformation characteristics of soil.

Engineers use various methods to analyze soil stress. One commonly used approach is the application of stress distribution theories, such as Boussinesq's or Westergaard's equations. These theories provide mathematical equations to calculate stress distribution at different depths and distances from a loaded area. By understanding the stress distribution, engineers can assess the potential for soil failure and design appropriate measures to mitigate risks.

Another important aspect of soil stress analysis is determining the bearing capacity of soil. Bearing capacity refers to the ability of soil to support the load of a structure without excessive settlement or shear failure. Engineers use different methods, including Terzaghi's bearing capacity theory, to estimate the maximum load that soil can withstand.

Soil stress analysis also plays a significant role in slope stability analysis. Slopes are common features in civil engineering projects, and understanding the stress distribution within a slope is crucial in assessing its stability. Engineers use techniques like limit equilibrium analysis and finite element modeling to evaluate the factors influencing slope stability, such as soil properties, slope geometry, and external loads.

In conclusion, soil stress analysis is a vital component of geotechnical engineering, particularly for civil engineers. It involves evaluating the distribution and magnitude of forces within soil to ensure the stability and safety of structures. By understanding concepts like effective stress, stress distribution theories, bearing capacity, and slope stability analysis, engineers can make informed decisions in designing and constructing various infrastructure projects.

Chapter 3: Site Investigation and Geotechnical Testing

Importance of Site Investigation in Geotechnical Engineering

Introduction:
Site investigation is a crucial aspect of geotechnical engineering in construction projects. It involves the thorough examination and analysis of the subsurface conditions at a construction site. This subchapter explores the significance of site investigation and its role in ensuring the success and safety of civil engineering projects.

Understanding Subsurface Conditions:
Site investigation provides engineers with valuable information about the geological and geotechnical characteristics of the site. This knowledge is essential for making informed decisions regarding the design, construction, and maintenance of structures. By studying the soil and rock properties, groundwater conditions, and potential hazards, engineers can develop appropriate strategies to mitigate risks and enhance the overall stability and performance of the project.

Risk Assessment and Mitigation:
Through site investigation, engineers can identify potential risks and hazards associated with the site. This includes assessing the soil's stability, determining the presence of underground water, evaluating the risk of landslides or sinkholes, and identifying the potential for liquefaction during earthquakes. By understanding these risks, engineers can devise mitigation measures, such as slope stabilization techniques, dewatering methods, or foundation design modifications, to ensure the safety of the structure and its surroundings.

Design   Optimization:
Site investigation plays a crucial role in optimizing the design of geotechnical structures. By examining the subsurface conditions, engineers can select appropriate foundation types, determine bearing capacities, and assess the suitability of different construction materials. This information helps in optimizing the design, reducing costs, and ensuring the long-term stability and durability of the structure.

Construction   Planning   and   Execution:
Site investigations provide engineers with critical data for planning and executing construction activities. By understanding the subsurface conditions, engineers can develop effective construction techniques, sequence activities, and select appropriate machinery and equipment. This knowledge helps in minimizing construction delays, controlling costs, and ensuring the timely completion of the project.

Environmental   Considerations:
Site investigation is also crucial for assessing the environmental impact of construction projects. It helps engineers identify potential contamination risks, evaluate the suitability of disposal sites, and develop appropriate measures for protecting the environment. By understanding the site's characteristics, engineers can implement sustainable construction practices and minimize the project's ecological footprint.

Conclusion:
In conclusion, site investigation is of utmost importance in geotechnical engineering and civil construction projects. It provides engineers with valuable information about subsurface conditions, helps in risk assessment and mitigation, optimizes design, facilitates construction planning and execution, and ensures environmental

sustainability. By investing in thorough site investigations, engineers can enhance the overall safety, performance, and success of their projects.

Types of Geotechnical Investigations

Geotechnical investigations are a crucial part of any construction project, especially in the field of civil engineering. These investigations provide engineers with valuable information about the properties of the soil and rock at a construction site, helping them make informed decisions and design safe and reliable structures. This subchapter will explore the various types of geotechnical investigations commonly carried out in civil engineering projects.

1. Desk Study:
Before any site investigation, engineers must conduct a desk study to gather existing information about the site. This includes reviewing geological maps, previous reports, and any available data on soil and rock properties. The desk study provides a foundation for further investigations and helps identify potential risks and challenges.

2. Field Reconnaissance:
A field reconnaissance involves visiting the construction site to visually inspect the area and note any visible features or geological conditions. This initial assessment allows engineers to gain a preliminary understanding of the site and identify potential areas of concern.

3. Geophysical Surveys:
Geophysical surveys use various techniques to investigate subsurface conditions without extensive excavation. Methods such as seismic surveys, ground-penetrating radar, and electrical resistivity are employed to map the subsurface layers and identify potential hazards like voids, buried structures, or variations in soil composition.

4. Borehole Logging:
Borehole logging involves drilling boreholes into the ground to obtain soil and rock samples. These samples are

then analyzed in a laboratory to determine their physical and mechanical properties. Borehole logging helps in understanding the stratigraphy of the site, identifying soil types, and assessing their engineering properties.

5. Cone Penetration Testing (CPT): CPT is a widely used in-situ testing method to determine soil properties. A cone penetrometer is pushed into the ground, and data is collected on the resistance encountered and pore pressure. This test provides valuable information on soil strength, settlement characteristics, and the presence of any weak or compressible layers.

6. Plate Load Tests: Plate load tests are performed to determine the bearing capacity and settlement characteristics of the soil. A large plate is loaded incrementally, and the settlement is measured to assess the soil's response to the applied load. This test helps engineers in designing appropriate foundations for structures.

7. Laboratory Testing: Laboratory testing involves analyzing soil and rock samples collected during the investigation. Various tests, including grain size analysis, compaction tests, consolidation tests, and shear strength tests, are conducted to determine the engineering properties of the materials. These laboratory results are used in the design and construction of foundations, retaining walls, and other structures.

In conclusion, geotechnical investigations play a vital role in civil engineering projects. By employing various investigation techniques, engineers can gather critical information about the site's soil and rock properties, enabling them to make informed decisions and design structures that are safe and sustainable.

Geotechnical Testing Techniques

Geotechnical testing techniques play a crucial role in the field of civil engineering, providing engineers with valuable insights into the properties and behavior of soils and rocks. These tests help engineers make informed decisions during the design and construction phases of infrastructure projects, ensuring their safety, stability, and durability.

One of the most common geotechnical testing techniques is the Standard Penetration Test (SPT), which involves driving a split-barrel sampler into the ground using a hammer and recording the number of blows required to penetrate a certain depth. The SPT provides engineers with information about soil resistance, which is used to estimate soil strength and determine foundation design parameters.

Another widely used testing technique is the Cone Penetration Test (CPT), which involves pushing a cone-shaped penetrometer into the ground and measuring the resistance to penetration. The CPT provides engineers with data on soil properties such as shear strength, compressibility, and permeability. It is especially useful in assessing the stability of slopes, embankments, and foundations.

In addition to these field tests, laboratory testing techniques are also employed to further analyze soil and rock samples. These tests include the determination of physical properties such as grain size distribution, moisture content, and specific gravity. Engineers also conduct tests to evaluate the engineering properties of soils, such as shear strength, consolidation characteristics, and permeability.

Geotechnical testing techniques have evolved over time, incorporating advanced technologies and equipment. For instance, the use of geophysical methods such as ground-penetrating radar and seismic surveys can provide engineers with information on subsurface conditions, helping them detect voids, fractures, and other potential hazards.

Furthermore, with the advent of computer modeling and numerical analysis, engineers can simulate various geotechnical scenarios and evaluate the behavior of soil and rock masses under different loading conditions. This allows for more accurate predictions and safer designs.

In conclusion, geotechnical testing techniques are vital tools for civil engineers in the field of geotechnical engineering. By providing valuable data on soil and rock properties, these tests enable engineers to make informed decisions during the design, construction, and maintenance of infrastructure projects. Continued advancements in testing techniques and technologies further enhance the accuracy and reliability of geotechnical investigations, ensuring the safety and success of construction projects.

Laboratory Testing of Soils

In the field of geotechnical engineering, laboratory testing of soils is an essential practice that helps engineers assess the properties and behavior of soil samples. This subchapter aims to provide engineers, specifically those in the civil engineering niche, with a comprehensive understanding of the various laboratory tests conducted on soils.

The primary purpose of laboratory testing is to analyze the physical, mechanical, and chemical properties of soils. By subjecting soil samples to controlled laboratory conditions, engineers can obtain accurate and reliable data to support the design and construction of structures such as buildings, bridges, and highways.

One of the most common laboratory tests is the classification of soils. This test categorizes soils into different classes based on their particle size distribution, enabling engineers to determine their engineering properties and potential behavior. Another crucial test is the compaction test, which measures the density and moisture content of soils to ensure their suitability for construction purposes.

Furthermore, laboratory tests also include the determination of the shear strength of soils. By conducting tests such as the direct shear test or the triaxial shear test, engineers can evaluate the stability of slopes, foundations, and retaining walls. These tests help in understanding how soils respond to external forces and provide valuable insights for designing safe and stable structures.

Laboratory testing also plays a significant role in assessing the permeability of soils, which is crucial in analyzing drainage systems and predicting soil behavior during rainfall events. By conducting tests like the constant head or falling head permeability tests, engineers can quantify the flow of water through soils and determine their suitability for specific applications.

Moreover, laboratory tests allow engineers to investigate the compressibility and consolidation characteristics of soils. By performing tests such as the oedometer test, engineers can determine the settlement potential of soils, which is vital for ensuring the stability and long-term performance of structures.

In conclusion, laboratory testing of soils is an integral part of geotechnical engineering, particularly in the civil engineering niche. These tests provide engineers with valuable data regarding the physical, mechanical, and chemical properties of soils, allowing them to make informed decisions during the design and construction of various structures. By understanding the principles and practices of laboratory testing, engineers can ensure the safety, durability, and efficiency of their geotechnical projects.

Field Testing of Soils

Field testing of soils is an essential aspect of geotechnical engineering in construction. Civil engineers must possess a thorough understanding of soil behavior and properties to ensure the safety, stability, and performance of various structures. Field testing provides engineers with valuable data that helps in the design and construction of foundations, retaining walls, embankments, and other geotechnical projects.

One of the most commonly used field tests is the Standard Penetration Test (SPT). This test measures the resistance of soil to penetration by a standard sampler driven by a hammer. The number of blows required for the sampler to penetrate specific depths provides engineers with information about the soil's strength and density. SPT results are used to estimate the bearing capacity of the soil, determine its shear strength parameters, and assess liquefaction potential.

Another widely used field test is the Cone Penetration Test (CPT). This test involves pushing a cone-shaped penetrometer into the soil at a constant rate. The cone's resistance and pore pressure measurements are recorded to evaluate soil properties such as shear strength, stiffness, and stratification. CPT results are particularly useful in assessing the settlement characteristics and liquefaction potential of a site.

Field tests also help engineers determine the soil's compaction characteristics. The Standard Proctor Test and Modified Proctor Test are commonly employed to evaluate soil compaction and moisture-density relationships. These tests involve compacting soil samples at various moisture contents and measuring their dry unit weight. The results assist in selecting suitable compaction methods for construction projects and ensuring the stability and performance of compacted soil layers.

In addition to these tests, engineers may conduct field vane shear tests to determine the undrained shear strength of cohesive soils, plate load tests to assess the bearing capacity of shallow foundations, and permeability tests to evaluate the soil's ability to allow water flow.

Field testing of soils is crucial for engineers to obtain accurate and reliable data about the ground conditions at a construction site. This data forms the basis for geotechnical engineering design and provides insights into soil behavior, which is essential for ensuring the safety and efficiency of civil engineering projects. By conducting appropriate field tests and interpreting the results, engineers can make informed decisions and mitigate potential risks associated with soil instability and inadequate foundation design.

Instrumentation and Monitoring in Geotechnical Engineering

In the field of geotechnical engineering, instrumentation and monitoring play a crucial role in ensuring the safety and stability of civil engineering projects. By measuring and analyzing various parameters, engineers can gain valuable insights into the behavior of the ground and structures, enabling them to make informed decisions and take appropriate measures to mitigate risks.

This subchapter aims to provide engineers in the field of civil engineering with a comprehensive understanding of the principles and practices of instrumentation and monitoring in geotechnical engineering. It covers a wide range of topics, from the selection of appropriate instruments to the interpretation of monitoring data.

The subchapter begins by introducing the fundamental concepts of instrumentation and monitoring, emphasizing their importance in assessing the performance of geotechnical structures. It highlights the need for accurate and reliable data to ensure the safety of construction projects and the importance of selecting appropriate instruments for different applications.

Next, the subchapter delves into the various types of instruments commonly used in geotechnical engineering. It discusses the principles behind instruments such as inclinometers, piezometers, settlement gauges, and strain gauges, and explains their applications in measuring parameters such as slope movements, pore water pressures, settlement, and deformation.

The subchapter also covers the installation and calibration of instruments, providing engineers with practical guidelines to ensure accurate measurements. It emphasizes the importance of proper instrument placement, calibration procedures, and regular maintenance to obtain reliable data throughout the lifespan of a project.

Furthermore, the subchapter discusses the different monitoring techniques used in geotechnical engineering. It explores the use of automated systems, remote sensing technologies, and real-time monitoring to enhance the efficiency and effectiveness of data collection and analysis.

Lastly, the subchapter addresses the interpretation and analysis of monitoring data. It explains the importance of data visualization, statistical analysis, and trend monitoring in identifying potential risks and taking appropriate actions. Case studies and examples are provided to illustrate the practical application of these techniques.

Overall, this subchapter aims to equip engineers in the field of civil engineering with the knowledge and skills necessary to effectively implement instrumentation and monitoring in geotechnical engineering projects. By understanding the principles and practices outlined in this subchapter, engineers can ensure the safety and success of their projects, while minimizing risks and maximizing efficiency.

Chapter 4: Foundation Engineering

Introduction to Foundation Engineering

Foundation engineering is a crucial aspect of civil engineering that plays a significant role in the design and construction of various structures. It involves the analysis and design of foundations to support and transmit the loads of a structure to the underlying soil or rock mass.

This subchapter aims to provide engineers in the field of civil engineering with a comprehensive introduction to foundation engineering. It explores the fundamental principles, practices, and considerations involved in designing safe and reliable foundations for different types of structures.

The subchapter begins by defining what foundation engineering is and why it is essential in construction projects. It emphasizes the significance of understanding the behavior of soil and rock materials, as well as the loads imposed on the foundation, to ensure the stability and safety of the structure.

Next, the subchapter delves into the different types of foundation systems commonly used in civil engineering. It discusses shallow foundations, such as spread footings and mat foundations, which are suitable for structures with relatively low loads and stable soil conditions. It also explores deep foundations, including piles and drilled shafts, which are employed when the soil near the surface is weak or unstable.

The subchapter then covers the various geotechnical investigations and tests carried out to determine the soil and rock properties at a construction site. It introduces engineers to techniques such as soil sampling, laboratory testing, and in-situ testing, which aid in gathering crucial data for the foundation design process.

Furthermore, the subchapter highlights the common factors that influence foundation design, such as the characteristics of the soil or rock, the magnitude and nature of the loads, and the environmental conditions. It emphasizes the importance of considering these factors to ensure the foundation's stability and prevent issues such as settlement, bearing capacity failure, or slope instability.

To conclude, this subchapter serves as a comprehensive introduction to foundation engineering for engineers in the field of civil engineering. It provides a solid foundation of knowledge and understanding of the principles and practices involved in designing safe and efficient foundations for various construction projects. By familiarizing themselves with these concepts, engineers can confidently undertake foundation design and contribute to the successful completion of infrastructure projects.

Types of Foundations

Foundations are essential elements of any civil engineering project, providing stability and support to structures. The choice of foundation type is crucial as it depends on various factors such as soil conditions, the load-bearing capacity of the soil, and the type of structure to be built. This subchapter will provide an overview of the different types of foundations commonly used in geotechnical engineering for civil construction projects.

1. Shallow Foundations: Shallow foundations, also known as spread footings, are a common choice for buildings with relatively low loads. They are designed to transfer the load to a shallow depth, typically within the upper layers of the soil. Shallow foundations include strip footings, which run along the length of a wall, and isolated footings, which support individual columns.

2. Deep Foundations: Deep foundations are used when the soil near the surface is unable to support the load. These foundations transfer the load to deeper, more competent soil layers or rock. Common types of deep foundations include driven piles, which are hammered or vibrated into the ground, and drilled shafts, which are constructed by excavating a hole and filling it with concrete or reinforcing steel.

3. Pile Foundations: Pile foundations are a type of deep foundation that consists of long, slender elements driven into the ground or formed in situ. They are used to transfer loads through weak or compressible soil layers to more competent strata at depth. Pile foundations can be made of timber, steel, or concrete and are suitable for a wide range of soil conditions.

4. Mat Foundations:
Mat foundations, also known as raft foundations, are used when the loads on a structure are too heavy for individual footings. They distribute the load of the structure over a large area of soil, reducing the stresses on the soil. Mat foundations are commonly used for high-rise buildings, industrial structures, and heavily loaded structures.

5. Floating Foundations:
Floating foundations, also called buoyant foundations, are used in areas with high water tables or areas subjected to frequent flooding. These foundations are designed to float on the water table and provide support to the structure without excessive settlement. Floating foundations are commonly used in coastal regions and areas with soft or saturated soils.

In conclusion, selecting the appropriate foundation type is crucial for ensuring the stability and durability of civil engineering structures. Factors such as soil conditions, load requirements, and environmental considerations play a significant role in determining the best foundation type for a particular project. Engineers must carefully evaluate these factors to ensure the successful construction of safe and reliable structures.

Shallow Foundations

In the world of civil engineering, the construction of buildings and structures relies heavily on the strength and stability of their foundations. Shallow foundations, also known as spread footings, are one of the most commonly used types of foundations in construction projects. This subchapter aims to provide engineers, specifically those in the field of civil engineering, with a comprehensive understanding of shallow foundations and their significance in geotechnical engineering.

Shallow foundations are designed to transfer the load of a structure to the underlying soils near the surface. They are typically used when the soil is strong enough to support the load without excessive settlement. This type of foundation is suitable for low-rise structures such as residential buildings, small commercial buildings, and light industrial structures. Understanding the principles and practices of designing and constructing shallow foundations is crucial for engineers involved in these types of projects.

This subchapter begins by delving into the fundamental concepts of geotechnical engineering and soil mechanics, providing engineers with a solid foundation of knowledge. It then explores the various types of shallow foundations, including individual footings, strip footings, and mat foundations. The subchapter discusses the factors that influence the selection of the appropriate type of shallow foundation, such as the nature of the soil, the load requirements of the structure, and the presence of nearby structures or utilities.

The subchapter also covers the design considerations and calculations involved in designing shallow foundations. It highlights important factors such as bearing capacity, settlement, and lateral stability, and provides engineers with practical guidance on how to assess and mitigate potential risks. Additionally, the subchapter addresses construction techniques and quality control measures to ensure the successful implementation of shallow foundations.

Throughout the subchapter, real-world case studies and examples are provided to illustrate the principles and practices discussed. These examples showcase the challenges faced by engineers in different geotechnical contexts and shed light on effective problem-solving approaches.

Overall, this subchapter on shallow foundations aims to equip engineers in the field of civil engineering with the necessary knowledge and skills to design and construct strong and stable foundations for various types of structures. By understanding the principles and practices of geotechnical engineering in relation to shallow foundations, engineers will be better prepared to ensure the safety and integrity of their construction projects.

Deep Foundations

Deep foundations play a crucial role in the field of geotechnical engineering, especially in civil engineering projects. This subchapter will delve into the principles and practices of deep foundations, providing engineers with a comprehensive understanding of their importance and applications in construction.

Deep foundations are structural elements that transfer loads from the superstructure to underlying layers of soil or rock. Unlike shallow foundations, which rely on the bearing capacity of surface soils, deep foundations penetrate deeper into the ground to reach stronger and more stable layers. These foundations are typically used when the soil near the surface is unable to support the loads imposed by the structure.

One of the most common types of deep foundations is the pile foundation. Piles are long, slender columns made of various materials, such as concrete, steel, or timber. They are driven or drilled into the ground until they reach a sufficient depth where the soil or rock can provide adequate support. Pile foundations are widely used in high-rise buildings, bridges, and other structures where significant loads need to be transferred to the ground.

This subchapter will explore the design considerations for deep foundations, including the selection of appropriate pile types, pile capacity calculations, and load testing methods. Engineers will learn about the different factors that influence the choice of pile types, such as soil conditions, structural requirements, and construction constraints.

Furthermore, the subchapter will cover the construction techniques employed for deep foundations, including pile driving, auger drilling, and hydraulic jacking. It will discuss the importance of quality control during construction, ensuring that the installed deep foundations meet the design specifications.

Additionally, the subchapter will address the various challenges and potential issues that engineers may encounter when working with deep foundations. These challenges may include soil liquefaction, lateral load resistance, and pile integrity concerns. Engineers will gain insights into the evaluation and mitigation strategies for such challenges to ensure the long-term stability and safety of the structures.

Overall, this subchapter on deep foundations will equip engineers in the field of civil engineering with the essential knowledge and tools to effectively design, construct, and evaluate deep foundations for a wide range of construction projects. With a strong understanding of deep foundations, engineers will be able to ensure the structural integrity and longevity of the built environment.

Design Considerations for Foundations

Foundations play a crucial role in the construction of any structure, providing the necessary support and stability to ensure the longevity and safety of the building. As engineers in the field of civil engineering, it is essential to understand the key design considerations when it comes to foundations.

Soil Properties: The first and foremost consideration in foundation design is understanding the soil properties. Engineers must evaluate the soil's bearing capacity, settlement characteristics, and soil stability to determine the appropriate foundation design. Conducting thorough geotechnical investigations and soil testing is crucial to gather accurate data for these assessments.

Load Analysis: Another critical aspect is conducting a comprehensive load analysis. Engineers must consider both the dead loads (self-weight of the structure) and live loads (occupancy, furniture, etc.) that the foundation will need to support. Additionally, any anticipated future loads, such as renovations or expansions, should also be taken into account to ensure the foundation design is adequate for the long term.

Foundation Types: Various foundation types are available, and selecting the most suitable one is vital. Common foundation types include shallow foundations (spread footings and mat foundations) and deep foundations (piles and drilled shafts). Each foundation type has its advantages and limitations, depending on the soil conditions, load requirements, and site constraints. Engineers must carefully evaluate these factors to determine the most appropriate foundation type for a given project.

Structural Compatibility: The foundation design must be compatible with the structural system of the building.

Engineers need to ensure that the interaction between the foundation and the superstructure is properly coordinated to avoid any potential conflicts or structural failures. This involves considering factors such as the transfer of loads, differential movements, and settlement distributions.

Construction Techniques: The construction techniques utilized during foundation installation are crucial for ensuring the integrity and performance of the foundation. Engineers must consider factors such as excavation methods, dewatering techniques, and proper compaction to prevent soil instability or settlement issues. Adequate quality control and monitoring during construction are also essential to verify that the foundation is constructed according to design specifications.

Environmental Considerations: Lastly, engineers should consider any environmental factors that may impact the foundation design. This includes evaluating the potential for soil erosion, slope stability, groundwater conditions, and seismic activity. Incorporating appropriate measures to mitigate these environmental risks is essential for ensuring the long-term stability and durability of the foundation.

In conclusion, foundation design is a critical aspect of civil engineering, requiring careful consideration of various factors. By understanding the soil properties, conducting thorough load analysis, selecting the appropriate foundation type, coordinating with the structural system, implementing proper construction techniques, and considering environmental factors, engineers can ensure the successful design and construction of foundations that provide the necessary support and stability for any structure.

Foundation Construction Techniques

Introduction:
In the field of civil engineering, the construction of foundations plays a crucial role in ensuring the stability, durability, and safety of structures. The foundation is the base on which the entire structure rests, transferring the loads from the superstructure to the underlying soil or rock. This subchapter on Foundation Construction Techniques aims to provide engineers with an overview of the fundamental principles and practices involved in constructing strong and reliable foundations.

Types of Foundations:
Before delving into the construction techniques, it is essential to understand the different types of foundations commonly used in civil engineering. This subchapter will cover shallow foundations, such as spread footings and mat foundations, as well as deep foundations, including pile foundations and drilled shafts. Each type has its own advantages and is suitable for specific soil conditions and structural requirements.

Site Investigation and Soil Analysis:
The success of foundation construction lies in the thorough understanding of the site's geotechnical characteristics. Engineers need to undertake comprehensive site investigations and soil analysis to assess the soil properties, bearing capacity, and potential risks such as liquefaction or settlement. This subchapter will provide guidelines on conducting site investigations and interpreting soil reports to make informed decisions during foundation construction.

Design Considerations:
Designing foundations involves considering various factors, including the structural loads, soil conditions, and

groundwater levels. This subchapter will discuss the principles of foundation design, including the selection of appropriate foundation types, determining the foundation size and depth, and incorporating factors of safety. It will also cover techniques for mitigating potential issues such as differential settlement or lateral soil movements.

Foundation Construction Techniques: The core focus of this subchapter lies in the construction techniques employed during foundation construction. It will discuss excavation methods, such as open excavation or caisson sinking, and the importance of proper dewatering techniques to control groundwater levels. Furthermore, it will cover the installation of different types of foundations, including the use of formworks, reinforcement, and concrete pouring techniques. Special attention will be given to quality control and inspection procedures during construction.

Case Studies and Best Practices: To provide practical insights, this subchapter will include case studies and examples of foundation construction projects from real-world applications. These case studies will highlight the challenges faced, the techniques employed, and the lessons learned. Best practices, recommendations, and industry standards will also be discussed to ensure the construction of safe and reliable foundations.

Conclusion:
The foundation construction techniques discussed in this subchapter are essential knowledge for civil engineers involved in the design and construction of structures. By mastering these techniques and understanding the underlying principles, engineers can ensure the successful implementation of foundations that withstand the test of time and provide a solid base for any structure.

Chapter 5: Retaining Structures

Introduction to Retaining Structures

Retaining structures play a crucial role in the field of civil engineering, providing support and stability to slopes, embankments, and other areas where soil or rock needs to be contained. These structures are designed to resist lateral pressure exerted by the soil or water, ensuring safety and preventing soil erosion or collapse. This subchapter aims to introduce engineers in the field of civil engineering to the fundamental concepts and principles of retaining structures.

The first section of this subchapter will provide an overview of the different types of retaining structures commonly used in geotechnical engineering. This will include gravity walls, cantilever walls, sheet pile walls, and anchored walls. Each type will be explained in terms of their structural components, construction techniques, and suitability for various soil conditions.

The next section will delve into the design considerations for retaining structures, emphasizing the importance of conducting thorough site investigations and geotechnical testing. Engineers will learn how to analyze soil properties, including cohesion, angle of internal friction, and consolidation characteristics, and how these properties affect the stability and design of retaining structures. Additionally, the role of water pressure and its impact on the stability of retaining structures will be discussed.

The subchapter will also cover the various methods used in the design of retaining structures, including limit equilibrium analysis, which is based on the principles of statics and is commonly employed in practice. Other methods such as numerical modeling and geotechnical software will also be introduced, highlighting their advantages and limitations.

Furthermore, the subchapter will address the construction techniques and best practices for retaining structures, focusing on the importance of proper compaction, drainage systems, and reinforcement techniques. Special considerations for designing retaining structures in challenging environments, such as earthquake-prone areas, will also be discussed.

To enhance the understanding of the material, the subchapter will include case studies featuring real-world examples of successful retaining structure projects. These case studies will showcase the practical application of the concepts and principles discussed earlier, providing engineers with valuable insights and lessons learned.

In conclusion, this subchapter serves as an essential introduction to retaining structures for engineers in the field of civil engineering. By familiarizing themselves with the different types, design considerations, construction techniques, and real-world examples, engineers will be equipped with the knowledge and skills necessary to design and construct safe and efficient retaining structures in various geotechnical conditions.

Types of Retaining Structures

In the field of geotechnical engineering, retaining structures play a crucial role in providing stability to slopes and preventing soil erosion. These structures are designed to withstand the lateral pressure exerted by the soil and other external forces. In this subchapter, we will explore the different types of retaining structures commonly used in civil engineering.

1. Gravity Walls: Gravity walls rely on their weight to resist the lateral pressure exerted by the soil. These walls are typically made from concrete or stone and are suitable for retaining relatively low heights. They are cost-effective and easy to construct, making them commonly used in various construction projects.

2. Cantilever Walls: Cantilever walls are constructed using reinforced concrete and are designed to resist soil pressure by using a horizontal base slab and a vertical stem. These walls are ideal for retaining moderate to high heights and can be found in many infrastructure projects such as highways and bridges.

3. Sheet Pile Walls: Sheet pile walls are made of steel, concrete, or timber and are driven vertically into the ground. They are commonly used in waterfront structures, such as docks and quay walls, where retaining soil is required. Sheet pile walls are effective in areas with limited space and can resist high lateral pressures.

4. Anchored Walls: Anchored walls are used when the lateral pressure exerted by the soil is too high for gravity or cantilever walls to withstand. These walls are reinforced with cables or rods that are anchored into the soil or rock behind the wall. Anchored walls are commonly used in deep excavations or where space constraints are present.

5. Gabion Walls: Gabion walls are constructed using wire mesh baskets filled with rocks or other suitable materials. These walls are flexible and allow for water drainage, making them ideal for areas with high water tables. Gabion walls are commonly used for erosion control, slope stabilization, and retaining walls in road construction.

6. Reinforced Earth Walls: Reinforced earth walls are constructed by embedding layers of metallic strips or grids into compacted backfill material. These walls provide excellent stability and are commonly used in highway projects, where high retaining heights and heavy loads are present.

In summary, retaining structures are essential in civil engineering to provide stability and prevent soil erosion. The choice of retaining structure depends on factors such as soil conditions, height requirements, and project constraints. By understanding the different types of retaining structures, engineers can make informed decisions when designing and constructing projects that require soil retention.

Design Considerations for Retaining Structures

Retaining structures play a critical role in civil engineering projects, providing stability and support to soil and other materials. They are commonly used to prevent soil erosion, control water flow, and create level surfaces in areas with varying topography. Designing a successful retaining structure requires careful consideration of various factors to ensure its long-term performance and safety. This subchapter aims to highlight key design considerations for engineers working in the field of geotechnical engineering, specifically in the niche of civil engineering.

One of the primary design considerations for retaining structures is the type of soil or material being retained. Different soils exhibit varying properties such as cohesion, angle of repose, and permeability. Understanding these characteristics is crucial in selecting the appropriate design approach, as well as determining the necessary reinforcement measures. Engineers must also consider the potential for soil movement, including settlement, consolidation, and lateral pressure, which can significantly impact the stability of the retaining structure.

Another important consideration is the hydrological conditions of the site. Engineers need to assess the water table level, surface runoff, and the potential for hydrostatic pressure buildup behind the retaining structure. These factors can lead to increased lateral pressures on the structure and potential failure. Proper drainage systems, including weep holes and subsurface drains, should be incorporated into the design to alleviate excess water pressure.

Structural design considerations involve selecting the appropriate type of retaining structure based on the specific project requirements. Common types include cantilever walls, gravity walls, anchored walls, and reinforced soil walls. Each type has its advantages and limitations, and the engineer must consider factors such as available space, material availability, construction costs, and aesthetics when making a decision.

The design must also account for external loads such as surcharge loads, seismic forces, and traffic loads. Engineers should conduct thorough calculations and analysis to determine the impact of these loads on the retaining structure. Adequate strength and stability should be ensured through the proper selection of materials, reinforcement, and connections.

Additionally, construction considerations must be taken into account during the design phase. Engineers should evaluate available construction techniques, equipment, and resources to ensure that the design can be effectively implemented. Construction sequencing, temporary shoring, and soil compaction requirements are important aspects to consider during the design process.

In conclusion, designing retaining structures requires careful consideration of various factors to ensure their long-term performance and safety. Soil characteristics, hydrological conditions, structural design, external loads, and construction considerations are all vital elements that engineers must address. By incorporating these considerations into the design process, engineers can create robust and efficient retaining structures that meet the specific needs of civil engineering projects.

Construction Methods for Retaining Structures

Retaining structures play a crucial role in civil engineering projects, providing support and stability to slopes and ensuring the safety and longevity of structures. This subchapter aims to provide engineers in the field of civil engineering with an overview of the various construction methods available for retaining structures.

1.	Gravity	Retaining	Walls: Gravity retaining walls rely on their own weight to resist the lateral pressure exerted by the soil. They are typically constructed using heavy materials such as concrete or natural stone, which provide stability and prevent soil movement. This method is widely used for low to medium height retaining structures, and engineers must consider factors such as wall height, soil properties, and water pressure when designing and constructing gravity retaining walls.

2.	Cantilever	Retaining	Walls: Cantilever retaining walls are structurally efficient and commonly used for higher retaining structures. They are designed with a base slab and a vertical stem, which relies on the principle of leverage to resist soil pressure. This method requires careful consideration of factors such as soil type, wall dimensions, and reinforcement detailing to ensure stability and prevent excessive deflection.

3.	Anchored	Retaining	Walls: Anchored retaining walls are suitable for projects where space is limited or where high lateral pressures are expected. This method involves the use of anchors or tiebacks, which are tensioned against the retained soil to provide additional stability. Anchored retaining walls require specialized construction techniques, including drilling, grouting, and stressing of the anchor system.

4. Sheet Pile Walls: Sheet pile walls are commonly used in temporary structures or where quick construction is required. This method involves driving interlocking sheet piles into the ground to create a barrier against soil movement. Sheet pile walls are often used in waterfront structures, underground parking lots, and deep excavations. Engineers must consider factors such as soil conditions, sheet pile material, and driving techniques to ensure efficient construction and long-term stability.

5. Gabion Walls: Gabion walls are constructed using wire mesh baskets filled with stones or other granular materials. They are flexible, cost-effective, and suitable for various soil conditions. Gabion walls are commonly used for erosion control, slope stabilization, and retaining walls in areas with limited access. Engineers must consider factors such as wall height, soil properties, and proper drainage to ensure the effectiveness of gabion walls.

In conclusion, the construction methods for retaining structures discussed in this subchapter provide engineers in the field of civil engineering with a comprehensive understanding of the different techniques available. Each method has its own advantages and limitations, and proper consideration of soil conditions, structural requirements, and construction techniques is essential to ensure the stability and durability of retaining structures.

Stability Analysis of Retaining Structures

In the field of civil engineering, the stability analysis of retaining structures plays a crucial role in ensuring the safety and long-term performance of various construction projects. Retaining structures are commonly used in the design of foundations, embankments, slopes, and other geotechnical engineering applications. Understanding and analyzing the stability of these structures is essential to prevent potential failures and ensure the integrity of the overall construction project.

This subchapter will delve into the principles and practices of stability analysis, providing engineers with a comprehensive understanding of the factors influencing the stability of retaining structures. It will cover various stability analysis methods and techniques, enabling engineers to make informed decisions and design robust retaining systems.

The subchapter will begin by introducing the fundamental concepts of soil mechanics, including soil properties, behavior, and the principles of soil-structure interaction. Engineers will gain insights into the different types of retaining structures, such as gravity walls, cantilever walls, and anchored walls, and their respective advantages and limitations.

Next, the subchapter will explore the key factors influencing the stability of retaining structures. These factors include soil properties, water content, slope geometry, surcharge loads, and seismic forces. By understanding the interplay between these factors, engineers will be able to identify potential failure mechanisms and develop appropriate design measures to mitigate risks.

The subchapter will then delve into various stability analysis methods, such as limit equilibrium methods, finite element analysis, and numerical modeling. Engineers will learn how to apply these methods to assess the stability of retaining structures under different loading conditions and to determine the required safety factors.

Additionally, the subchapter will address the importance of ongoing monitoring and maintenance of retaining structures. Engineers will understand the significance of regular inspections and the use of instrumentation to detect any signs of distress or potential instability. This proactive approach allows for timely remedial measures to be implemented, ensuring the long-term stability and safety of the retaining structures.

By the end of this subchapter, engineers specializing in civil engineering and geotechnical engineering will have a comprehensive understanding of stability analysis for retaining structures. Armed with this knowledge, they will be able to design and construct robust retaining systems that withstand the test of time and ensure the safety of the built environment.

Chapter 6: Slope Stability Analysis

Introduction to Slope Stability Analysis

Slope stability analysis is a crucial aspect of geotechnical engineering, specifically in the field of civil engineering. It involves the assessment of the stability of natural and man-made slopes, such as embankments, hillsides, and excavations, to ensure the safety and integrity of structures built on or adjacent to these slopes.

In this subchapter, we will provide engineers with a comprehensive introduction to slope stability analysis, covering the fundamental principles and practices involved in assessing and mitigating the risks associated with unstable slopes.

The first section of this subchapter will delve into the importance of slope stability analysis in civil engineering projects. Engineers will learn about the potential consequences of slope failures, including property damage, loss of life, and environmental impacts. By understanding these risks, engineers can appreciate the significance of conducting thorough slope stability analysis to prevent such disasters.

The subsequent section will introduce engineers to the various factors that influence slope stability. These factors include geological and geotechnical properties of the soil or rock mass, groundwater conditions, slope geometry, and external forces. Engineers will gain insights into how these factors interact and contribute to the overall stability of a slope.

Next, the subchapter will explore the different methods and techniques used in slope stability analysis. Engineers will be introduced to both analytical and numerical methods, such as limit equilibrium analysis, finite element analysis, and slope stability charts. The advantages and limitations of each method will be discussed, enabling engineers to select the most appropriate technique based on the specific project requirements.

Furthermore, engineers will learn about key parameters and variables involved in slope stability analysis, such as shear strength, slope angle, pore water pressure, and safety factors. Understanding these parameters is essential for accurate and reliable stability assessments.

The subchapter will conclude with a discussion on slope stabilization measures. Engineers will be introduced to various techniques for stabilizing slopes, including soil reinforcement, drainage systems, retaining walls, and slope grading. Case studies and practical examples will be provided to illustrate the application of these measures in real-world scenarios.

By the end of this subchapter, engineers will have a solid foundation in slope stability analysis and be equipped with the knowledge and tools necessary to identify potential slope stability issues and implement appropriate mitigation measures.

Factors Affecting Slope Stability

In the field of geotechnical engineering, understanding the factors that affect slope stability is crucial for engineers involved in civil engineering projects. Slope stability refers to the ability of a natural or man-made slope to resist the forces that may cause it to fail or collapse. Failure of a slope can have serious consequences, including property damage, environmental impact, and even loss of life. Therefore, engineers must thoroughly analyze and consider various factors when designing and constructing slopes.

One of the primary factors affecting slope stability is the geological and geotechnical properties of the materials comprising the slope. The type of soil or rock, its strength, cohesion, and permeability, as well as the presence of any water, play a vital role in determining the stability of the slope. For instance, clayey soils with high water content are more prone to instability than well-drained sandy soils. Engineers must evaluate the soil and rock properties through site investigations, laboratory testing, and geotechnical analysis to assess the potential risks associated with slope failure.

Another crucial factor is the slope geometry and profile. The height, angle, and shape of a slope significantly influence its stability. Steep slopes are generally more susceptible to failure, especially if the soil or rock is weak. Engineers must consider the impact of the slope angle on stability and design appropriate measures, such as reinforcement techniques or slope flattening, to mitigate potential risks.

External forces acting on the slope can also affect its stability. These forces include seismic activity, rainfall, water pressure, and human activities like excavation and construction. Seismic events, such as earthquakes, can generate ground shaking that may reduce the strength and cohesion of the soil, increasing the likelihood of slope failure. Similarly, heavy rainfall can cause an increase in pore water pressure, leading to decreased soil strength and potential instability. Engineers must assess these external forces and incorporate appropriate design measures, such as drainage systems or slope stabilization techniques, to enhance slope stability.

Furthermore, the presence of vegetation and the overall climate of the region can impact slope stability. Plant roots can help bind the soil, providing additional stability to the slope. Conversely, deforestation or removal of vegetation can increase the risk of slope failure. Climate conditions, such as freeze-thaw cycles or prolonged rainfall, can also affect the stability of slopes.

In conclusion, engineers involved in civil engineering projects must consider several factors when analyzing and designing slopes to ensure their stability. The geological and geotechnical properties of the soil and rock, slope geometry, external forces, vegetation, and climate all play a significant role in determining slope stability. By thoroughly evaluating these factors and implementing appropriate design measures, engineers can minimize the risk of slope failure and ensure the safety and longevity of the constructed structures.

Types of Slope Failures

In the field of geotechnical engineering, understanding the various types of slope failures is crucial for engineers, particularly those specializing in civil engineering. Slope failures can result in devastating consequences, including property damage, loss of life, and disruption of infrastructure. This subchapter will explore the different types of slope failures, their causes, and the methods used to prevent or mitigate them.

1. Slides: Slope slides occur when there is a failure along a distinct surface, resulting in the movement of a mass of soil or rock. This type of failure is typically caused by excessive water saturation, changes in pore pressure, geological discontinuities, or inadequate slope design. Engineers employ various techniques such as stabilization measures, surface drainage, and slope reinforcement to prevent or mitigate slope slides.

2. Flows: Slope flows involve the movement of saturated or partially saturated soil or rock as a viscous fluid. Common examples include debris flows, mudflows, and avalanches. Flows are often triggered by heavy rainfall, earthquakes, or rapid snowmelt. Engineers employ measures like dewatering, retaining walls, and diversion channels to prevent or control slope flows.

3. Falls: Slope falls refer to the detachment and free fall of material from a slope due to gravity. Rockfalls and soil falls are the two main categories of slope falls. They can be caused by weathering, erosion, seismic activity, or human-induced factors such as excavation. Engineers use techniques like rock bolting, slope mesh, and slope stabilization measures to mitigate the risk of slope falls.

4. Creep: Slope creep is a slow, continuous movement of soil or rock downslope. It occurs over an extended period

due to factors such as the expansion and contraction of soil, freeze-thaw cycles, or the presence of weak layers. Engineers employ methods like soil nailing, ground anchors, and terracing to minimize the effects of creep and ensure slope stability.

5. Subsidence: Slope subsidence involves the gradual sinking or settling of the ground surface due to a collapse or compaction of underlying materials. It can be caused by natural processes like dissolution of soluble rocks or human activities such as mining or groundwater extraction. Engineers use techniques like grouting, ground improvement, and monitoring systems to prevent or control subsidence.

By understanding the various types of slope failures and their causes, engineers can develop appropriate design and construction strategies to ensure the stability and safety of slopes in civil engineering projects. It is essential to consider site-specific conditions, geological factors, and the potential risks associated with each type of slope failure. Regular monitoring and maintenance are also crucial to detect early signs of slope instability and prevent catastrophic failures in the long run.

Slope Stability Analysis Methods

In the field of geotechnical engineering, the stability of slopes is of utmost importance for engineers involved in civil engineering projects. Slope failures can lead to disastrous consequences, such as landslides, soil erosion, and damage to infrastructure. Therefore, it is crucial for engineers to employ reliable slope stability analysis methods to ensure the safety and stability of slopes.

There are several methods available for analyzing slope stability, each with its own advantages and limitations. This subchapter aims to provide engineers in the field of civil engineering with an overview of these methods, enabling them to make informed decisions and design stable slopes.

One commonly used method is the limit equilibrium analysis, which assumes that the slope fails along a potential slip surface. This method involves calculating the forces acting on the slope and comparing them to the strength properties of the soil. Various techniques, such as Bishop's method, Spencer's method, and Janbu's method, are employed to determine the factor of safety, which indicates the stability of the slope.

Another approach is the finite element method, which involves dividing the slope into small elements and analyzing the behavior of each element under various loading conditions. This method is particularly useful for complex slope geometries and heterogeneous soil conditions. It provides a detailed understanding of the stress and deformation distribution within the slope, allowing engineers to make accurate predictions of slope stability.

Furthermore, probabilistic methods, such as the Monte Carlo simulation and the First Order Second Moment method, consider the uncertainties associated with soil properties and external factors, such as rainfall and seismic events. These methods provide engineers with a probabilistic assessment of slope stability, taking into account the variability in soil properties and loading conditions.

In addition to the aforementioned methods, this subchapter will also discuss other slope stability analysis techniques, including the shear strength reduction method, the method of slices, and the numerical modeling approach. Each method will be presented with its underlying principles, assumptions, and limitations.

By understanding and applying these slope stability analysis methods, engineers in the field of civil engineering can effectively evaluate the stability of slopes, design appropriate slope reinforcement measures, and mitigate potential slope failures. This subchapter aims to equip engineers with the necessary knowledge and tools to ensure the stability and safety of slopes in geotechnical engineering projects.

Mitigation Measures for Slope Stability

In the field of civil engineering, ensuring the stability of slopes is of paramount importance to prevent potential disasters such as landslides and slope failures. Slopes are susceptible to various factors, including geological conditions, weather conditions, and human activities, which can compromise their stability. Therefore, it is crucial for engineers to implement effective mitigation measures to minimize the risks associated with slopes.

One of the primary mitigation measures for slope stability is proper site investigation and geological assessment. Engineers must thoroughly understand the geological conditions of the site, including soil types, rock formations, and groundwater levels. This information helps in identifying potential areas of concern and allows engineers to develop suitable design solutions.

To enhance slope stability, engineers often employ techniques such as slope reinforcement. This involves the use of geotechnical materials such as geotextiles, geogrids, and geosynthetic anchors, which provide additional tensile strength to the soil mass. By reinforcing the slope, the potential for soil movement and failure is significantly reduced.

Drainage systems are another crucial mitigation measure for slope stability. Excessive water accumulation can significantly weaken the soil, leading to slope instability. Engineers must design and implement proper drainage systems, including surface drains, subsurface drains, and slope stabilization ditches, to divert water away from the slope and maintain its stability.

Vegetation plays a vital role in slope stability as well. Engineers often incorporate bioengineering techniques, such as the planting of vegetation, to enhance slope stability. The roots of plants help bind the soil particles together, providing additional resistance against erosion and slope failure. Vegetation also helps in reducing surface water runoff, thereby minimizing the chances of slope instability.

In certain cases, where the slope is highly susceptible to failure, engineers may resort to slope modification techniques. This includes slope flattening, benching, or constructing retaining structures like retaining walls and soil nails. These modifications redistribute the forces acting on the slope, reducing the potential for slope failure.

Regular monitoring of slopes is imperative to ensure their stability over time. Engineers should install monitoring devices such as inclinometers, piezometers, and settlement gauges to measure any changes in slope behavior. This enables early detection of any potential instability, allowing engineers to take necessary preventive measures promptly.

In conclusion, mitigating slope stability is a critical aspect of geotechnical engineering in the field of civil engineering. By employing techniques such as proper site investigation, slope reinforcement, drainage systems, vegetation planting, slope modification, and regular monitoring, engineers can effectively minimize the risks associated with slope instability. These measures ensure the safe and sustainable construction of infrastructure in areas prone to slope failures, promoting the overall stability and longevity of the built environment.

Chapter 7: Earthquake Engineering and Geotechnical Considerations

Introduction to Earthquake Engineering

Earthquakes are natural disasters that can cause significant damage to infrastructure, leading to loss of life and economic consequences. As engineers, it is crucial to understand the principles and practices of earthquake engineering to design and construct safe and resilient structures in earthquake-prone regions. This subchapter provides an introduction to earthquake engineering, focusing on its relevance to civil engineers.

Earthquake engineering is a specialized field within civil engineering that deals with the study of seismic forces and their effects on structures. It involves the analysis, design, and construction of buildings, bridges, dams, and other infrastructure to withstand earthquake-induced ground shaking, ground rupture, and other seismic hazards.

The fundamental concept in earthquake engineering is the recognition that earthquakes are inevitable, and their occurrence can be characterized by various parameters, including magnitude, frequency, and ground motion characteristics. Understanding these parameters is essential to assess the seismic hazard and develop design guidelines for structures.

One of the key considerations in earthquake engineering is the ground response analysis, which involves evaluating how the ground motion affects the behavior of structures. This analysis helps engineers determine the design parameters, such as the seismic design forces and the expected displacement, acceleration, and velocity responses of structures.

The seismic design of structures involves incorporating various measures to mitigate the potential damage caused by earthquakes. These measures include designing buildings with adequate lateral strength and stiffness, implementing proper foundation systems to minimize ground shaking effects, and using materials and techniques that can absorb and dissipate seismic energy.

Furthermore, earthquake engineering also addresses the importance of site selection and soil-structure interaction. The characteristics of the site, such as soil type, liquefaction potential, and slope stability, significantly influence the behavior of structures during earthquakes. Understanding the interaction between the structure and the underlying soil is crucial to ensure the safety and stability of the infrastructure.

In conclusion, earthquake engineering plays a vital role in the field of civil engineering, as it enables engineers to design structures that can withstand the forces generated by earthquakes. By incorporating seismic design principles and practices, engineers can mitigate the potential damage caused by earthquakes, safeguarding lives and reducing economic losses. This subchapter serves as an introduction to earthquake engineering, providing engineers with essential knowledge to tackle the challenges associated with seismic hazards in construction projects.

Seismic Hazards and Site Characterization

In the field of civil engineering, understanding seismic hazards and conducting site characterization studies are crucial for ensuring the safety and integrity of structures and infrastructure in earthquake-prone areas. This subchapter will delve into the fundamental concepts and practices related to seismic hazards and site characterization, providing engineers with the necessary knowledge to assess and mitigate risks effectively.

Seismic hazards refer to the potential dangers posed by earthquakes, including ground shaking, ground rupture, and secondary effects such as liquefaction and landslides. Engineers need to comprehend the various factors that contribute to seismic hazards, such as tectonic plate movement, fault lines, and historical seismic activity in a region. By analyzing seismic hazard maps and data, engineers can determine the level of risk associated with a specific site and develop appropriate design strategies to withstand potential earthquakes.

Site characterization plays a crucial role in assessing the seismic hazards of a particular location. It involves gathering information about the geologic, geotechnical, and seismic properties of the site. Engineers employ various methods, including geophysical surveys, borehole drilling, laboratory testing, and seismic refraction studies, to obtain accurate data. This information is then used to determine the site's soil and rock properties, including their strength, stiffness, and liquefaction potential, which are essential for designing earthquake-resistant structures.

The subchapter will discuss the different techniques and tools used in site characterization, emphasizing their significance in assessing seismic hazards. It will cover topics such as soil sampling and testing, geotechnical instrumentation, and seismic hazard analysis methods. Additionally, the subchapter will explore the importance of considering site-specific factors, such as topography, soil liquefaction potential, and the presence of active faults, in evaluating seismic hazards.

Engineers must also be aware of relevant codes, regulations, and design standards related to seismic hazards. The subchapter will provide an overview of these guidelines, ensuring that engineers are familiar with the necessary requirements for designing structures in earthquake-prone areas. It will emphasize the importance of incorporating seismic design principles into the construction process to enhance the resilience of buildings and infrastructure.

By comprehending seismic hazards and conducting thorough site characterization studies, engineers can make informed decisions during the design and construction phases. This subchapter aims to equip civil engineers with the knowledge and skills necessary to assess seismic risks accurately and develop robust and safe structures that can withstand earthquakes.

Geotechnical Considerations in Earthquake Engineering

Earthquakes pose significant challenges to civil engineers, especially when it comes to designing and constructing structures that can withstand their destructive forces. Understanding the geotechnical aspects of earthquake engineering is crucial for engineers working in the field of civil engineering. This subchapter aims to provide engineers with a comprehensive overview of the geotechnical considerations involved in earthquake engineering.

One of the key factors in earthquake engineering is the response of the ground to seismic waves. Engineers must analyze the site-specific seismic hazard to determine the potential ground motion that a structure may experience during an earthquake. This involves studying the geological and geotechnical characteristics of the site, including soil type, groundwater conditions, and the presence of fault lines. By understanding these factors, engineers can make informed decisions regarding the design and construction of structures that can withstand ground shaking.

Soil liquefaction is another critical geotechnical consideration in earthquake engineering. During an earthquake, loose or saturated soils can lose their strength and behave like a liquid, leading to the sinking or tilting of structures. Engineers must assess the susceptibility of soils to liquefaction and take appropriate measures to mitigate its effects. This may involve ground improvement techniques such as compaction, grouting, or the installation of reinforcement elements to increase soil stability.

The design of foundations is another crucial aspect of geotechnical considerations in earthquake engineering. Engineers must ensure that the foundations of structures are capable of withstanding the dynamic loads imposed by seismic events. This may involve the use of specialized foundation systems, such as pile foundations, which can provide increased stability and resistance to ground shaking.

In addition to these considerations, engineers must also address the potential for landslides and slope instability during earthquakes. Steep slopes, weak soils, and heavy rainfall can increase the likelihood of landslides, which can pose significant risks to infrastructure and human lives. Geotechnical investigations and slope stability analysis are essential in determining the appropriate slope reinforcement and stabilization measures.

Overall, geotechnical considerations play a vital role in earthquake engineering. By understanding the geological and geotechnical aspects of a site, engineers can design and construct structures that are resilient to seismic forces. This subchapter aims to provide engineers in the field of civil engineering with an in-depth understanding of these considerations, equipping them with the knowledge and tools necessary to ensure the safety and stability of structures in earthquake-prone areas.

Seismic Design of Foundations and Retaining Structures

In the field of civil engineering, the seismic design of foundations and retaining structures is of paramount importance. As engineers, it is crucial for us to understand the principles and practices involved in ensuring the safety and stability of structures in areas prone to earthquakes. This subchapter delves into the intricacies of seismic design, providing engineers with the necessary knowledge to tackle the challenges associated with geotechnical engineering in seismic zones.

Foundations play a vital role in transferring loads from the superstructure to the underlying soil. However, in seismic regions, the dynamic forces generated by earthquakes can pose a significant threat to the stability of these foundations. Therefore, it is imperative to design foundations that can withstand these forces and prevent the structure from experiencing excessive deformations or collapse.

The subchapter explores various types of foundation systems suitable for seismic regions, such as shallow foundations, deep foundations, and pile foundations. It discusses the factors that influence the selection of an appropriate foundation system, including soil properties, site conditions, and the anticipated seismic forces. Additionally, the subchapter provides engineers with design guidelines for ensuring the adequate strength and stiffness of foundations under seismic loading.

Retaining structures, including walls and slopes, also require careful consideration in seismic design. The subchapter delves into the analysis and design of these structures to withstand the lateral forces induced by earthquakes. It covers topics such as soil-structure interaction, seismic earth pressures, and the use of reinforcement techniques to enhance the stability and performance of retaining structures.

Furthermore, the subchapter addresses the importance of site characterization and geotechnical investigations in seismic design. Engineers are guided on how to assess the seismic hazard at a particular site, including the determination of design ground motions and the estimation of soil liquefaction potential. These assessments form the basis for developing appropriate design criteria and selecting suitable construction techniques to mitigate the effects of seismic forces.

Throughout the subchapter, practical examples and case studies are provided to illustrate the application of seismic design principles in real-world scenarios. These examples not only demonstrate the challenges faced by engineers but also showcase innovative solutions and best practices.

In conclusion, the seismic design of foundations and retaining structures is a critical aspect of geotechnical engineering in construction. This subchapter equips engineers in the field of civil engineering with the necessary tools and knowledge to ensure the safety and stability of structures in seismic regions. By adhering to the principles and practices outlined in this subchapter, engineers can confidently design and construct structures that can withstand the forces unleashed by earthquakes, ultimately protecting lives and property.

Seismic Analysis and Design Methods

In the field of civil engineering, seismic analysis and design methods play a crucial role in ensuring the safety and stability of structures in areas prone to earthquakes. Understanding the behavior of soils and structures under seismic loading is essential for engineers involved in geotechnical engineering. This subchapter aims to provide a comprehensive overview of the principles and practices related to seismic analysis and design, specifically tailored for engineers in the niche of civil engineering.

The first section of this subchapter delves into the fundamentals of seismic analysis. It starts by explaining the different types of seismic waves and their characteristics, emphasizing the importance of understanding their impact on structures. The subchapter then explores various methods used to analyze the dynamic response of structures subjected to seismic forces, including the response spectrum analysis, time history analysis, and equivalent static analysis.

The subsequent section focuses on the design of structures to withstand seismic loads. It discusses the concept of seismic design criteria, highlighting the importance of performance-based design approaches to ensure the reliability and safety of structures. The subchapter further addresses the design considerations for different types of geotechnical structures, such as shallow and deep foundations, retaining walls, and slopes, under seismic conditions. It emphasizes the role of soil-structure interaction and the use of appropriate design codes and guidelines.

Furthermore, this subchapter explores advanced topics in seismic analysis and design. It covers the seismic assessment and retrofitting of existing structures, including evaluation techniques, strengthening methods, and the use of innovative materials and technologies. The subchapter also provides an overview of the latest developments in the field, such as performance-based seismic design, seismic isolation, and base isolation systems.

Throughout the subchapter, practical examples, case studies, and illustrations are included to enhance the understanding of seismic analysis and design methods. These examples highlight real-world applications and demonstrate the challenges faced by engineers in geotechnical engineering.

By the end of this subchapter, engineers in the field of civil engineering will have a solid foundation in seismic analysis and design methods. They will be equipped with the knowledge and tools necessary to assess seismic hazards, analyze the dynamic response of structures, and design structures that can withstand the forces generated by earthquakes. This subchapter serves as a valuable resource for engineers seeking to enhance their expertise in geotechnical engineering and contribute to the safe and sustainable construction of structures in seismic-prone regions.

Chapter 8: Ground Improvement Techniques

Introduction to Ground Improvement Techniques

Ground improvement techniques play a crucial role in the field of geotechnical engineering, especially in the construction industry. This subchapter aims to provide engineers, particularly those in the field of civil engineering, with a comprehensive introduction to various ground improvement techniques commonly employed in construction projects. By understanding these techniques, engineers can effectively address challenges related to soil stabilization, foundation support, and slope stability, among others.

The subchapter will begin by defining ground improvement and its significance in construction. It will highlight the importance of undertaking ground improvement measures to enhance soil properties, increase bearing capacity, and reduce settlement. The subchapter will emphasize the need for engineers to thoroughly assess soil conditions and determine the most suitable ground improvement technique for a given project.

The content will then delve into different types of ground improvement techniques, starting with mechanical techniques. This section will cover methods such as compaction, vibro-compaction, and dynamic compaction, which aim to densify loose or weak soils. The subchapter will explain the principles and applications of each technique, along with their advantages and limitations.

The subsequent section will focus on chemical ground improvement techniques. These techniques involve the use of chemical additives to alter soil properties, such as cement stabilization, lime stabilization, and soil grouting.

The content will provide engineers with insights into the mechanisms of these techniques and their effectiveness in improving soil characteristics.

The subchapter will further explore biological ground improvement techniques, including biogrouting and bio-stabilization. These techniques utilize microbial or plant-based processes to enhance soil properties, such as strength and permeability. The content will discuss the benefits and limitations of these techniques and their application in specific geotechnical scenarios.

Lastly, the subchapter will discuss geosynthetic-based ground improvement techniques, such as soil reinforcement using geotextiles, geogrids, and geocells. It will explain the principles behind these techniques and their role in stabilizing slopes, reinforcing foundations, and mitigating settlement.

By the end of this subchapter, engineers will have a solid understanding of various ground improvement techniques and their applications in civil engineering projects. This knowledge will enable them to make informed decisions regarding the selection and implementation of appropriate ground improvement measures, ensuring the safety, stability, and longevity of construction projects.

Soil Stabilization Methods

Soil stabilization methods are crucial in the field of geotechnical engineering, particularly in the realm of civil engineering projects. These methods play a vital role in enhancing the properties of soil, making it more suitable for construction purposes. This subchapter aims to provide engineers with an in-depth understanding of various soil stabilization techniques commonly used in construction projects.

One of the primary methods of soil stabilization is the addition of chemicals. This technique involves the introduction of additives to improve soil properties such as strength, durability, and cohesion. Chemical agents like lime, cement, and fly ash are commonly used to stabilize soil. Engineers must understand the chemical reactions between these additives and the soil to achieve effective stabilization.

Another prominent technique in soil stabilization is mechanical stabilization. This method involves altering the physical characteristics of the soil to enhance its engineering properties. Techniques such as compaction, soil replacement, and soil densification are commonly employed. Engineers must carefully assess the soil's natural properties and determine the most suitable mechanical stabilization technique for the specific construction project.

Furthermore, soil stabilization can also be achieved through the application of geosynthetics. Geosynthetics are synthetic materials specifically designed for use in geotechnical engineering projects. They can be used to reinforce soil, provide drainage, and control erosion. Engineers must be knowledgeable about the different types of geosynthetics available, including geotextiles, geogrids, and geomembranes, and their appropriate applications.

Additionally, this subchapter will cover innovative soil stabilization methods, such as the use of bio-polymers and microbial-induced calcite precipitation (MICP). Bio-polymers, derived from renewable sources, have shown promising results in stabilizing soil. MICP, on the other hand, involves the injection of bacteria into the soil, which precipitates calcium carbonate, thereby increasing soil strength.

Overall, the subchapter on soil stabilization methods will equip engineers in the field of civil engineering with a comprehensive understanding of the techniques available for enhancing soil properties. By mastering these methods, engineers will be able to make informed decisions when selecting the most suitable soil stabilization technique for their construction projects.

Soil Improvement Techniques

In the field of civil engineering, soil improvement techniques play a crucial role in ensuring the stability and durability of structures. Engineers are often faced with the challenge of building on sites with weak or unsuitable soils, which can pose significant risks to the integrity of the construction. Therefore, it becomes imperative to employ various soil improvement techniques to enhance the engineering properties of the soil and mitigate potential issues.

One commonly used soil improvement technique is compaction. By compacting the soil, engineers increase its density and reduce its susceptibility to settlement. This technique is particularly useful for loose or sandy soils. Compaction can be achieved through various methods, such as dynamic compaction, vibro-compaction, or the use of heavy rolling equipment. The choice of method depends on the site conditions and the desired level of compaction.

Another widely employed technique is soil stabilization. This process involves altering the soil's physical or chemical properties to enhance its strength and reduce its susceptibility to erosion or deformation. Common stabilization methods include the addition of cement, lime, or fly ash to the soil, which chemically react with the soil particles to improve their engineering characteristics. Additionally, soil stabilization can be achieved through techniques like soil mixing, where cement or other binders are mechanically mixed with the soil to enhance its strength and stability.

In certain cases, soil reinforcement techniques are necessary to improve the soil's load-bearing capacity. These techniques involve the introduction of materials, such as geosynthetics, into the soil to increase its tensile strength and resistance to deformation. Geosynthetics, such as geotextiles, geogrids, or geocells, are commonly used to reinforce slopes, embankments, or retaining walls, providing additional stability and preventing soil erosion.

Lastly, soil drainage techniques are essential for improving the soil's permeability and reducing the risk of water-related issues. By installing drainage systems, engineers can effectively control the amount of water in the soil, preventing excessive saturation and potential instability. Techniques like the installation of drain pipes, French drains, or the use of permeable materials can significantly enhance the soil's drainage properties.

In conclusion, soil improvement techniques are fundamental for engineers in the field of civil engineering to ensure the stability and longevity of structures. Compaction, soil stabilization, soil reinforcement, and soil drainage are among the key techniques used to enhance the engineering properties of soil and mitigate potential risks. By implementing these techniques, engineers can overcome the challenges posed by weak or unsuitable soils, ultimately leading to safe and durable construction projects.

Deep Soil Mixing

Deep soil mixing is a widely used technique in geotechnical engineering that involves the mechanical mixing of soil with cementitious materials to improve its engineering properties. This process is typically employed in civil engineering projects where weak or unstable soil conditions are encountered, and a strong foundation is required.

The deep soil mixing method involves the use of specialized equipment that combines soil and cementitious materials in situ to create a homogenous soil-cement mixture. The equipment consists of a mixing tool, usually a rotating auger, that is lowered into the ground to the desired depth. As the auger rotates, it breaks up the existing soil and mixes it with the cementitious materials, resulting in a uniformly mixed soil-cement column.

One of the key advantages of deep soil mixing is its ability to treat large volumes of soil quickly and efficiently. This method is particularly useful in areas with limited access or in congested urban environments where traditional excavation and replacement methods may not be feasible. Additionally, deep soil mixing can be used to improve soil properties such as strength, stiffness, and permeability, making it an ideal solution for foundation support and ground improvement.

In civil engineering, deep soil mixing has been successfully applied in various projects, including the construction of highways, bridges, buildings, and underground structures. It has proven to be an effective and cost-efficient method for stabilizing soft or loose soils, reducing settlement, and increasing the bearing capacity of the soil.

Engineers involved in civil engineering projects can benefit from understanding the principles and practices of deep soil mixing. By incorporating this technique into their design and construction plans, engineers can ensure the stability and longevity of their structures while minimizing the need for costly foundation modifications or remedial measures.

In conclusion, deep soil mixing is a valuable technique in geotechnical engineering that offers numerous benefits for civil engineering projects. Its ability to improve soil properties and provide a strong foundation makes it an indispensable tool for engineers working in the field of geotechnical engineering. By incorporating deep soil mixing into their projects, engineers can enhance the performance and durability of their structures, ultimately leading to safer and more efficient construction practices.

Grouting and Injection Techniques

Grouting and Injection Techniques in Geotechnical Engineering

In the field of geotechnical engineering, grouting and injection techniques play a crucial role in the construction and stabilization of civil engineering projects. These techniques involve the injection of a fluid material, known as grout, into the ground or structures to improve their stability, prevent water seepage, and strengthen the soil or rock mass.

Grouting is primarily used to fill voids, fractures, and cavities in the ground or structures. It can be performed both above and below ground level, depending on the specific requirements of the project. The grout material used can vary, but commonly includes cement, bentonite, or chemical grouts. The selection of the grout material depends on factors such as the desired strength, permeability, and setting time.

Injection techniques, on the other hand, involve the injection of grout into the ground to improve its load-bearing capacity or to control groundwater flow. This technique is often used in projects such as foundation stabilization, soil consolidation, and waterproofing of tunnels or underground structures.

One commonly used grouting technique is permeation grouting, which involves injecting low-viscosity grout into the soil or rock mass. This technique helps in filling the voids and improving the soil's strength and stiffness. Another widely employed technique is compaction grouting, which uses high-viscosity grout to densify loose or weak soils. It involves injecting the grout at high pressures to fracture the soil and compact it.

In civil engineering projects, grouting and injection techniques are extensively used for various applications. For instance, in the construction of tunnels, grouting is crucial for ensuring the stability of the surrounding soil or rock mass. It helps prevent water ingress and strengthens the ground to withstand the tunneling forces.

Moreover, grouting and injection techniques are also essential for foundation stabilization. By injecting grout into the soil beneath a foundation, engineers can improve its load-bearing capacity and prevent settlement or subsidence issues. This technique is particularly useful in areas with weak or compressible soils.

In summary, grouting and injection techniques are critical components of geotechnical engineering in civil construction projects. These techniques enable engineers to improve the stability, strength, and waterproofing of the ground and structures. By employing various grout materials and injection methods, engineers can tackle geotechnical challenges and ensure the long-term durability and safety of the constructed facilities.

Geosynthetics and Reinforcement Methods

In the field of civil engineering, geosynthetics and reinforcement methods play a crucial role in enhancing the stability and performance of geotechnical structures. Geosynthetics refer to synthetic materials used in geotechnical applications to improve soil behavior, enhance drainage, control erosion, and provide overall reinforcement. These materials, such as geotextiles, geogrids, geomembranes, and geocomposites, offer engineers a versatile and cost-effective solution for various construction challenges.

Geosynthetics are commonly used in applications such as soil stabilization, embankment construction, retaining walls, reinforced slopes, and landfills. The selection of the appropriate geosynthetic material depends on the specific requirements of the project, including soil type, load conditions, and environmental factors.

One of the most widely used geosynthetics is geotextiles. These permeable fabrics are designed to separate different soil layers, prevent the mixing of fine and coarse materials, and enhance the overall stability of the structure. Geotextiles act as filters, allowing water to pass through while retaining soil particles, thus preventing erosion and maintaining the long-term performance of the project.

Geogrids, on the other hand, are high-strength materials used for soil reinforcement. They are typically made of polymers and are characterized by their open-grid structure. Geogrids provide tensile strength to the soil, reducing deformation and increasing the load-bearing capacity of the structure. They are commonly used in reinforced soil walls, steep slopes, and paved roads, where they distribute the applied loads and minimize the potential for settlement or failure.

Geomembranes are impermeable geosynthetics used for environmental containment applications. They are often utilized in landfill liners, reservoirs, and wastewater treatment facilities to prevent the migration of contaminants into the surrounding soil and groundwater. Geomembranes are highly resistant to chemical degradation and provide an effective barrier against seepage and leakage.

Lastly, geocomposites combine multiple geosynthetic materials to provide a combined function. For example, a geocomposite may consist of a geotextile layer for filtration, a geogrid for reinforcement, and a geomembrane for containment. This integrated approach allows engineers to optimize the performance of geotechnical structures while reducing material usage and installation time.

In summary, geosynthetics and reinforcement methods are indispensable tools in modern civil engineering. By incorporating these materials into geotechnical projects, engineers can enhance stability, improve drainage, control erosion, and ensure the long-term performance of structures. Understanding the properties and applications of geosynthetics is essential for engineers involved in the design, construction, and maintenance of geotechnical projects.

Chapter 9: Geotechnical Aspects of Construction Materials

Geotechnical Properties of Construction Materials

In the field of civil engineering, understanding the geotechnical properties of construction materials is crucial for ensuring the stability, durability, and overall performance of structures. These properties provide engineers with valuable insights into the behavior and characteristics of various materials used in construction, allowing them to make informed decisions during the design and construction process.

One of the primary geotechnical properties that engineers need to consider is the soil's composition and classification. Different types of soils, such as clay, silt, sand, and gravel, exhibit distinct properties that directly impact the behavior of structures built on them. By understanding the soil composition, engineers can assess factors like shear strength, compaction characteristics, permeability, and settlement potential, which are essential for determining appropriate foundation designs and construction techniques.

Shear strength is another critical geotechnical property that engineers must evaluate. It refers to the soil's ability to resist shear forces, which can cause the material to deform or fail. By conducting shear strength tests, engineers can determine the maximum load a soil can withstand without yielding, allowing them to design foundations, retaining walls, and slopes with adequate safety factors.

The permeability of construction materials is also a vital consideration. Permeability refers to the ability of a material to allow the flow of fluids, such as water or gases, through its pores or voids. Understanding the permeability of soils is crucial for managing water drainage, preventing seepage, and protecting structures from potential damage caused by hydrostatic pressure or soil liquefaction.

Furthermore, engineers must assess the compaction characteristics of construction materials. Compaction is the process of reducing the voids in soil by applying mechanical energy, which increases its density and improves its load-bearing capacity. By understanding the compaction characteristics of soils, engineers can ensure that the materials used in construction are adequately compacted to meet specified engineering requirements.

Other geotechnical properties that engineers need to consider include the bearing capacity of soils, their settlement potential, and the potential for soil erosion. Each of these properties plays a crucial role in determining the overall stability and safety of structures.

In conclusion, understanding the geotechnical properties of construction materials is essential for civil engineers involved in the design and construction of structures. By evaluating the composition, shear strength, permeability, compaction characteristics, and other relevant properties, engineers can make informed decisions regarding foundation design, construction techniques, and overall structural integrity. These considerations are vital for ensuring the long-term stability, durability, and performance of civil engineering projects.

Geotechnical Considerations for Concrete

Concrete is one of the most widely used construction materials in civil engineering projects. It provides both structural support and durability to various types of structures, including buildings, bridges, and dams. However, the successful implementation of concrete structures heavily relies on geotechnical considerations. Engineers working in the field of civil engineering must carefully assess the geotechnical factors that can impact the performance and longevity of concrete structures.

Understanding the soil conditions is crucial when designing and constructing concrete structures. Engineers must evaluate the soil's bearing capacity, settlement characteristics, and soil composition to ensure the stability and integrity of the structure. Geotechnical investigations, including soil testing and site surveys, help engineers determine the appropriate foundation design and construction methods for concrete structures.

One of the key geotechnical considerations for concrete is the presence of expansive soils. These soils have the potential to undergo significant volume changes due to moisture fluctuations. Expansive soils can exert a substantial amount of pressure on concrete foundations, leading to cracks and structural damage. Engineers must identify these soils and implement appropriate measures, such as soil stabilization techniques or foundation designs that can accommodate the soil's expansive nature.

Another important geotechnical consideration is the presence of groundwater. High water tables or seepage can adversely affect the durability of concrete structures. Engineers must assess the groundwater conditions and incorporate appropriate measures to prevent water ingress into the concrete. This may involve the use of waterproofing membranes, drainage systems, or the selection of concrete mixes with low permeability.

Site-specific geotechnical considerations also play a critical role in the construction of concrete structures. Engineers must evaluate the seismic activity, slope stability, and potential for landslides or soil liquefaction in the project area. These factors can influence the design and reinforcement requirements for concrete structures to ensure their resistance to natural hazards.

In summary, geotechnical considerations are essential for the successful implementation of concrete structures. Engineers must thoroughly assess the soil conditions, including expansive soils and groundwater levels, to design and construct durable and stable concrete structures. Additionally, site-specific geotechnical factors, such as seismic activity and slope stability, must be carefully evaluated to ensure the safety and resilience of concrete structures in various environmental conditions. By incorporating geotechnical principles into concrete design and construction, engineers can enhance the performance and longevity of civil engineering projects.

Geotechnical Considerations for Asphalt

In the world of civil engineering, geotechnical considerations play a crucial role in the successful design and construction of various infrastructure projects, including roads and highways. One such aspect that engineers must carefully evaluate is the geotechnical properties and behavior of asphalt, a widely used material in road construction.

Asphalt, also known as bitumen, is a versatile material that provides flexibility, durability, and resistance to heavy loads and environmental factors. However, its performance largely depends on the underlying soil conditions and the interaction between the asphalt layer and the supporting subgrade.

When considering the geotechnical aspects of asphalt, engineers must evaluate several key factors. One of the primary considerations is the soil classification and its bearing capacity. The soil must be capable of supporting the weight of the asphalt layer, as well as the anticipated traffic loads. Engineers must conduct thorough soil investigations, including soil sampling and testing, to determine the soil's strength, compaction characteristics, and potential for settlement.

In addition to soil properties, engineers must also assess the moisture content and groundwater conditions. Excessive moisture can significantly weaken the subgrade and lead to stability issues, such as rutting and pavement failure. Proper drainage systems, including sub-surface drainage and adequate pavement slope, should be incorporated to prevent the accumulation of water within the pavement layers.

Furthermore, the geotechnical considerations for asphalt should also encompass the evaluation of slope stability.

Roads are often constructed on hilly terrains or slopes, which require a comprehensive understanding of the soil's shear strength and stability. Engineers must analyze the slope stability through methods such as slope stability analysis, soil shear strength testing, and geotechnical instrumentation to ensure the safety and longevity of the road.

Lastly, the subchapter on geotechnical considerations for asphalt should highlight the importance of proper compaction during construction. Adequate compaction of the asphalt layer is crucial to achieve the desired density, preventing the formation of voids and ensuring the required strength. Engineers must carefully monitor and control the compaction process using suitable equipment and techniques.

In summary, geotechnical considerations for asphalt are fundamental to the successful design and construction of roads and highways. Engineers must thoroughly evaluate the soil properties, moisture content, groundwater conditions, slope stability, and compaction requirements to ensure the longevity and performance of the asphalt pavement. By addressing these geotechnical aspects, civil engineers can effectively optimize the design, construction, and maintenance of asphalt roads, ultimately providing safe and reliable transportation infrastructure for communities.

Geotechnical Considerations for Aggregates

In the field of civil engineering, the use of aggregates is indispensable in construction projects. Aggregates, which are composed of various mineral materials such as sand, gravel, crushed stone, and slag, play a crucial role in providing stability, strength, and durability to different types of structures. However, before utilizing aggregates in construction, it is essential for engineers to consider various geotechnical aspects to ensure the safety and performance of the project.

One of the primary geotechnical considerations for aggregates is their suitability for the intended application. Engineers need to assess the physical and mechanical properties of aggregates to determine their compatibility with the project requirements. These properties include particle size distribution, shape, angularity, porosity, density, and strength. By evaluating these characteristics, engineers can select aggregates that will provide the necessary stability and load-bearing capacity for the specific construction task.

Another crucial geotechnical consideration is the potential for settlement and compaction of aggregates. Engineers need to understand how aggregates interact with the surrounding soil and how they can impact the overall settlement behavior of the project. Factors such as particle shape, gradation, and compaction energy can influence the settlement patterns, which can lead to differential settlement and potential structural damage. By considering these geotechnical aspects, engineers can design appropriate measures to mitigate settlement issues and ensure the long-term stability of the structure.

Additionally, engineers must also consider the geotechnical implications of using recycled aggregates. With the increasing emphasis on sustainable construction practices, the use of recycled aggregates has gained popularity. However, engineers need to evaluate the geotechnical properties of recycled aggregates, such as their strength, durability, and potential for contamination. Understanding these aspects is crucial to ensure that recycled aggregates meet the required standards and do not compromise the overall performance and safety of the construction project.

In conclusion, geotechnical considerations play a significant role in the successful utilization of aggregates in construction projects. Engineers need to thoroughly evaluate the suitability of aggregates, assess their potential for settlement and compaction, and consider the geotechnical implications of using recycled materials. By addressing these aspects, engineers can make informed decisions regarding the selection and use of aggregates, ensuring the stability, strength, and durability of the structures they design.

Geotechnical Considerations for Geosynthetics

Geosynthetics, which are synthetic materials used in geotechnical engineering applications, play a crucial role in enhancing the performance and longevity of civil engineering structures. This subchapter focuses on the geotechnical considerations that engineers need to keep in mind when incorporating geosynthetics in construction projects.

One of the key considerations is the site-specific soil conditions. Geosynthetics are typically used to improve the soil's mechanical properties, such as its strength, stability, and drainage characteristics. Therefore, engineers must thoroughly assess the soil's composition, its engineering properties, and its potential for settlement, erosion, and other geotechnical issues. This knowledge enables engineers to select the appropriate type and configuration of geosynthetics that will effectively address the specific soil challenges at hand.

Additionally, engineers must consider the design and installation of geosynthetics. The subchapter provides detailed guidelines on how to design geosynthetic-reinforced structures, such as retaining walls, embankments, and slopes. It covers aspects such as selecting the appropriate geosynthetic material, determining the required strength and stiffness parameters, and calculating the required reinforcement length and spacing. Furthermore, the subchapter emphasizes the importance of proper installation techniques, including adequate overlap and anchoring of geosynthetics, to ensure optimal performance and durability.

Another essential consideration is the long-term performance and durability of geosynthetics. Civil engineering structures are designed to have a long service life, and geosynthetics should be able to withstand the environmental and loading conditions over that period. The subchapter discusses key factors that can affect the durability of geosynthetics, such as exposure to harsh chemicals, ultraviolet radiation, temperature variations, and biological degradation. Engineers will learn about the various laboratory and field tests available to assess the durability of geosynthetics and ensure their suitability for the intended application.

Lastly, the subchapter addresses the economic considerations of incorporating geosynthetics in construction projects. While geosynthetics can offer significant benefits, such as cost savings, improved performance, and reduced environmental impact, their selection and implementation should be economically viable. Engineers will gain insights into the life-cycle cost analysis and value engineering techniques that can help optimize the use of geosynthetics while considering the project's budget and long-term maintenance costs.

Overall, this subchapter on geotechnical considerations for geosynthetics equips civil engineers with the knowledge and tools necessary to effectively incorporate these materials into their projects. By understanding the soil conditions, design and installation techniques, durability requirements, and economic considerations, engineers can make informed decisions that lead to safer, more efficient, and sustainable construction practices.

Chapter 10: Geotechnical Risk Assessment and Management

Introduction to Geotechnical Risk Assessment

Geotechnical risk assessment is a crucial aspect of civil engineering that helps engineers identify and mitigate potential hazards and uncertainties associated with the soil and rock materials underlying construction projects. By understanding and analyzing the risks involved in geotechnical engineering, engineers can make informed decisions, design effective foundations, and ensure the safety and stability of structures.

This subchapter provides an overview of geotechnical risk assessment, its importance in civil engineering, and the key principles and practices involved in the process. It aims to equip engineers with the necessary knowledge and tools to assess and manage geotechnical risks effectively.

The subchapter begins by highlighting the significance of geotechnical risk assessment in construction projects. As engineers, it is essential to recognize that the properties and behavior of soil and rock materials can vary significantly, introducing uncertainties and potential hazards. Geotechnical risk assessment enables engineers to identify these risks early on, assess their potential consequences, and develop appropriate strategies to mitigate them.

Next, the subchapter delves into the fundamental principles and methodologies of geotechnical risk assessment. It explores the different types of risks encountered in civil engineering, such as slope instability, liquefaction, and settlement, and describes the tools and techniques used to evaluate these risks. From site investigations and laboratory testing to numerical modeling and probabilistic analysis, engineers are introduced to the range of approaches available for assessing geotechnical risks.

Furthermore, the subchapter emphasizes the importance of considering both technical and non-technical factors in geotechnical risk assessment. Engineers must not only evaluate the physical characteristics of the soil but also take into account factors such as project specifications, environmental conditions, and regulatory requirements. By adopting a holistic approach, engineers can ensure that all relevant aspects of geotechnical risk are adequately addressed.

To illustrate the practical application of geotechnical risk assessment, the subchapter includes case studies and examples from real-world projects. These examples showcase how engineers have successfully identified and managed geotechnical risks, highlighting the value of a systematic and comprehensive risk assessment process.

In conclusion, this subchapter provides engineers in the field of civil engineering with a comprehensive introduction to geotechnical risk assessment. By understanding the principles and practices involved in assessing and managing geotechnical risks, engineers can enhance the safety, efficiency, and longevity of construction projects. Ultimately, this knowledge empowers engineers to make informed decisions and design structures that can withstand the challenges posed by the geotechnical environment.

Identification and Assessment of Geotechnical Risks

In the field of civil engineering, geotechnical risks play a crucial role in determining the success or failure of a construction project. Understanding the identification and assessment of these risks is essential for engineers involved in geotechnical engineering. This subchapter aims to provide a comprehensive overview of the process of identifying and assessing geotechnical risks, equipping engineers with the necessary knowledge and tools to mitigate potential challenges.

The first step in managing geotechnical risks is to identify potential hazards. This involves a thorough site investigation, where engineers analyze the geological conditions, soil properties, and groundwater levels. By understanding the site-specific characteristics, engineers can identify potential risks such as slope instability, soil liquefaction, or groundwater seepage. Furthermore, historical data on past projects in the same region can provide valuable insights into potential risks and their likelihood of occurrence.

Once the risks are identified, engineers must assess their potential impact on the project. This involves evaluating the consequences of failure, including potential damage to structures, environmental impacts, and safety hazards. By quantifying the consequences, engineers can prioritize risks and allocate resources accordingly.

Risk assessment also involves evaluating the likelihood of occurrence for each identified risk. This can be done through various methods, such as statistical analysis, numerical modeling, or expert judgment. By assigning probabilities to different risks, engineers can estimate their overall impact on the project's success.

It is important to note that geotechnical risks are highly uncertain and can vary throughout the project lifecycle. Therefore, engineers must continuously monitor and update risk assessments as new information becomes available. This iterative process allows for proactive risk management and reduces the likelihood of unexpected failures.

To aid engineers in the identification and assessment of geotechnical risks, numerous tools and techniques are available. These include geotechnical software for numerical modeling, remote sensing technologies for monitoring ground movements, and advanced laboratory testing methods for soil characterization. By utilizing these tools, engineers can enhance their understanding of geotechnical risks and make informed decisions to mitigate potential challenges.

In conclusion, the identification and assessment of geotechnical risks are vital components of successful civil engineering projects. By understanding the site-specific conditions, evaluating consequences and probabilities, and utilizing advanced tools, engineers can effectively manage geotechnical risks. This subchapter provides a comprehensive guide for engineers involved in geotechnical engineering, enabling them to navigate potential challenges and ensure the safe and efficient completion of construction projects.

Geotechnical Risk Mitigation Strategies

In the field of civil engineering, geotechnical risk mitigation strategies play a crucial role in ensuring the safety and stability of construction projects. These strategies aim to identify potential geotechnical hazards and implement measures to minimize or eliminate their impact on the project.

One of the primary geotechnical risk mitigation strategies is conducting thorough site investigations and geotechnical assessments. Engineers must assess soil properties, groundwater conditions, and geologic features to identify potential risks such as unstable slopes, liquefaction, or settlement. This information helps in designing appropriate foundations, retaining structures, and slope stabilization measures.

Another strategy is to implement effective slope stability measures. Slope failures can lead to significant damage and even loss of life, making it crucial to address this risk. Engineers can employ techniques such as slope stabilization using soil nails, ground anchors, or geosynthetic materials. In some cases, the installation of retaining walls or slope terracing may be necessary to ensure stability.

To mitigate the risks associated with liquefaction, engineers can implement various methods. These include soil improvement techniques such as deep soil compaction, dynamic compaction, or vibro-replacement. Installing vertical drains or using ground improvement materials like lime or cement can also help in reducing the risk of liquefaction during seismic events.

In addition to these measures, proper groundwater management is essential to mitigate geotechnical risks. Engineers must design and implement effective drainage systems to control groundwater levels and prevent potential issues such as soil erosion, settlement, or destabilization of slopes.

It is also crucial to consider the long-term effects of climate change on geotechnical stability. Rising sea levels, increased rainfall, and extreme weather events can significantly impact soil behavior. Engineers must incorporate climate change adaptation strategies into their designs, such as raising foundation elevations, implementing erosion control measures, or designing for increased loads.

Furthermore, continuous monitoring and risk assessment during and after construction are vital to identify and address any unforeseen geotechnical risks promptly. Monitoring techniques such as inclinometers, piezometers, or settlement gauges can be used to detect any signs of instability or changes in soil behavior.

In conclusion, geotechnical risk mitigation strategies are fundamental in ensuring the safety, stability, and longevity of civil engineering projects. By conducting thorough site investigations, implementing effective slope stability measures, managing groundwater, considering climate change, and implementing continuous monitoring, engineers can successfully mitigate potential geotechnical risks and ensure the success of their construction projects.

Geotechnical Risk Management in Construction Projects

Risk management is a critical aspect of any construction project, particularly when it comes to geotechnical engineering. The subchapter on Geotechnical Risk Management in Construction Projects aims to provide engineers, specifically in the field of civil engineering, with a comprehensive understanding of the principles and practices involved in mitigating geotechnical risks.

The subchapter begins by emphasizing the importance of geotechnical risk management in construction projects. It highlights the potential consequences of not addressing geotechnical risks effectively, such as project delays, cost overruns, and even structural failures. Engineers are encouraged to adopt a proactive approach to risk management, understanding that it is an ongoing process throughout the entire project lifecycle.

The content then delves into the key components of geotechnical risk management. It emphasizes the significance of a thorough site investigation, including geological and geotechnical assessments, to identify potential hazards and risks. Engineers are guided on how to interpret and analyze the data collected during site investigations, enabling them to make informed decisions regarding risk mitigation strategies.

The subchapter also explores various geotechnical risk assessment techniques and tools available to engineers. It delves into the concepts of probability and consequence analysis, discussing how these methods can aid in assessing and prioritizing risks. Additionally, engineers are introduced to various geotechnical software and simulation tools that can assist in evaluating and managing risks effectively.

Furthermore, the content covers risk mitigation strategies specific to geotechnical engineering. It discusses the importance of implementing suitable design measures, such as slope stabilization, retaining walls, and ground improvement techniques. Additionally, it emphasizes the significance of construction monitoring and quality control measures to ensure the effectiveness of risk mitigation measures throughout the construction process.

Throughout the subchapter, case studies and real-life examples are provided to illustrate the application of geotechnical risk management principles in different construction projects. These examples help to reinforce the importance of proactive risk management and provide engineers with practical insights on how to overcome geotechnical challenges effectively.

By the end of the subchapter, engineers in the niche of civil engineering will have gained a solid understanding of geotechnical risk management in construction projects. They will be equipped with the knowledge and tools necessary to identify, assess, and mitigate geotechnical risks, ultimately ensuring the successful and safe completion of construction projects.

Case Studies on Geotechnical Risk Assessment and Management

In the field of civil engineering, geotechnical risk assessment and management play a critical role in ensuring the safety and stability of construction projects. Understanding the potential risks associated with the ground conditions is essential for engineers to make informed decisions and implement effective mitigation measures. This subchapter presents a collection of case studies that showcase the importance of geotechnical risk assessment and management in various construction projects.

The first case study explores a high-rise building project situated in an area prone to seismic activity. The engineers faced the challenge of designing a foundation system that could withstand potential ground shaking. Through a comprehensive geotechnical risk assessment, they evaluated the site-specific seismic hazard, soil properties, and potential liquefaction susceptibility. By considering these factors, the engineers were able to design a foundation system that incorporated innovative techniques such as base isolation and soil improvement methods, ensuring the safety of the structure during seismic events.

The second case study focuses on a major infrastructure project involving the construction of a highway in a region with highly variable soil conditions. The engineers encountered significant challenges related to slope stability and soil settlement. They conducted extensive geotechnical investigations, including geophysical surveys and laboratory testing, to assess the soil properties and identify potential risks. By implementing appropriate slope stabilization measures and using ground improvement techniques, the engineers successfully mitigated the geotechnical risks and ensured the long-term stability of the highway.

Another case study examines a coastal development project that aimed to construct a marina and waterfront facilities. The engineers faced geotechnical challenges related to the presence of soft marine clay and potential soil liquefaction during seismic events. Through advanced geotechnical risk assessment techniques, including cone penetration testing and seismic hazard analysis, the engineers were able to develop a comprehensive risk management plan. This plan included the installation of deep foundation systems, ground improvement methods, and appropriate drainage systems to mitigate the geotechnical risks and ensure the stability of the structures.

These case studies highlight the significance of geotechnical risk assessment and management in civil engineering projects. By understanding the ground conditions and assessing potential risks, engineers can make informed decisions and implement effective mitigation measures to ensure the safety and stability of structures. Geotechnical risk assessment and management should be an integral part of every construction project, as it ultimately contributes to the overall success and longevity of the infrastructure.

Chapter 11: Sustainability in Geotechnical Engineering

Introduction to Sustainability in Geotechnical Engineering

Sustainability has become a critical aspect of engineering practices, particularly in the field of geotechnical engineering. As civil engineers, it is our responsibility to design and construct infrastructure that not only meets the immediate needs of society but also ensures the long-term viability of our projects. In this subchapter, we will explore the concept of sustainability in the context of geotechnical engineering and discuss its importance in civil engineering practices.

Geotechnical engineering focuses on understanding the behavior of soil and rock materials and using this knowledge to design foundations, slopes, and earth structures. Traditionally, the focus has been on ensuring the stability and safety of these structures. However, with the growing recognition of environmental concerns and the need for sustainable development, engineers must now consider the environmental, social, and economic impacts of their projects.

Sustainability in geotechnical engineering involves adopting practices that minimize the depletion of natural resources, reduce the generation of waste, and mitigate the environmental impact of construction activities. By incorporating sustainable principles, engineers can contribute to the preservation of natural ecosystems, reduce carbon emissions, and promote social well-being.

One of the key aspects of sustainability in geotechnical engineering is the use of environmentally friendly materials and techniques. For example, engineers can explore the use of alternative materials, such as recycled aggregates or geosynthetics, in place of traditional construction materials. Additionally, techniques like soil stabilization using bio-based additives or geotechnical reinforcement using renewable materials can help reduce the environmental impact of construction projects.

Another important consideration in sustainable geotechnical engineering is the management of resources. This includes optimizing the use of materials, reducing energy consumption during construction, and minimizing waste generation. By designing efficient foundation systems and implementing appropriate construction methods, engineers can significantly reduce the environmental footprint of their projects.

Furthermore, sustainable geotechnical engineering also involves considering the long-term performance and resilience of structures in the face of climate change. With the increasing frequency of extreme weather events, it is crucial to design infrastructure that can withstand such challenges and continue to function effectively. This may include considering the effects of climate change on soil properties, designing for increased flood resilience, or implementing measures to mitigate the effects of soil erosion.

In conclusion, sustainability is an integral part of geotechnical engineering in the field of civil engineering. By adopting sustainable practices, engineers can contribute to the preservation of the environment, promote social well-being, and ensure the long-term viability of their projects. This subchapter will delve deeper into the various aspects of sustainability in geotechnical engineering, providing engineers with the knowledge and tools necessary to incorporate sustainable principles into their design and construction practices.

Environmental Considerations in Geotechnical Engineering

As civil engineers, it is imperative that we consider the environmental impact of our projects throughout their lifecycle. Geotechnical engineering, which deals with the behavior of earth materials and their interaction with structures, is no exception. This subchapter explores the various environmental considerations that engineers must take into account when conducting geotechnical work.

First and foremost, soil erosion is a significant concern in geotechnical engineering projects. During construction, the removal of vegetation and disturbance of soil can result in increased erosion rates. This can lead to sedimentation in nearby bodies of water, causing water pollution and harming aquatic ecosystems. Engineers must implement erosion control measures such as the use of erosion control blankets, sediment barriers, and sedimentation ponds to mitigate these impacts.

Another crucial environmental consideration is the preservation of natural habitats. Many geotechnical projects involve land development, which can result in the destruction or fragmentation of habitats. This can have detrimental effects on local flora and fauna. Engineers should aim to minimize habitat disturbance by carefully planning project layouts and incorporating ecological corridors to maintain connectivity between habitat patches.

Furthermore, the disposal of excavated materials during geotechnical projects can have environmental implications. It is crucial to properly manage and dispose of these materials to prevent pollution and contamination of soil and water resources. Engineers should follow guidelines and regulations for waste management, including the proper classification, treatment, and disposal of excavated materials.

The choice of construction materials in geotechnical engineering also plays a significant role in environmental considerations. Engineers should consider the use of sustainable materials, such as recycled aggregates or alternative cementitious materials, that minimize the environmental footprint of the project. Additionally, the use of geosynthetics and other innovative techniques can help reduce the consumption of natural resources.

Lastly, geotechnical engineers should be mindful of the long-term effects of their projects on the environment. This includes considering the potential for soil subsidence, slope instability, and groundwater contamination. By conducting thorough site investigations and applying appropriate design methodologies, engineers can minimize these risks and ensure the long-term stability and sustainability of the project.

In conclusion, environmental considerations are crucial in geotechnical engineering to minimize the impact on natural ecosystems and resources. By implementing erosion control measures, preserving natural habitats, managing excavated materials, using sustainable construction materials, and considering long-term effects, engineers can contribute to a more environmentally conscious and sustainable approach to geotechnical projects.

Social and Economic Aspects in Geotechnical Engineering

In the field of geotechnical engineering, it is crucial for engineers to consider not only the technical aspects of construction projects but also the social and economic factors that can significantly impact the success of their designs. This subchapter explores the importance of understanding and incorporating social and economic aspects into geotechnical engineering practices, specifically within the niche of civil engineering.

Civil engineers play a vital role in building infrastructure that supports communities and contributes to economic growth. However, the success of these projects hinges on the ability to navigate the complex interplay between engineering solutions and their social and economic implications. By considering these aspects, engineers can ensure that their designs not only meet technical requirements but also align with the needs and aspirations of the society they serve.

One key social aspect to consider is the impact of construction projects on local communities. Engineers must assess the potential disruptions and inconveniences caused by construction activities, such as noise, dust, and traffic congestion. By implementing effective mitigation measures and engaging in open communication with the community, engineers can minimize the negative impacts and build positive relationships with stakeholders.

Additionally, understanding the economic aspects of geotechnical engineering is vital for project success. Engineers must evaluate the cost-effectiveness of different design options, considering factors such as construction materials, land acquisition, and maintenance requirements. By conducting thorough cost-benefit analyses, engineers can optimize project budgets and ensure that resources are allocated efficiently.

Furthermore, economic considerations extend beyond the design phase. Engineers must also assess the long-term economic sustainability of their projects. This involves evaluating the potential economic benefits that the infrastructure will bring to the community, such as increased property values, job creation, and improved transportation networks. By demonstrating the economic value of their designs, engineers can garner support from investors and policymakers, facilitating project implementation.

In conclusion, social and economic aspects play a crucial role in geotechnical engineering, particularly within the niche of civil engineering. By considering the social impacts of construction projects and incorporating economic evaluations into their designs, engineers can ensure that their work is not only technically sound but also responsive to the needs and aspirations of the community. This holistic approach to geotechnical engineering not only enhances project success but also fosters sustainable development and positive societal outcomes.

Sustainable Geotechnical Design and Construction Practices

In the field of civil engineering, sustainable design and construction practices have gained significant importance due to the growing concern for the environment and the need for long-term solutions. Geotechnical engineering, which deals with the behavior of soil and rock materials, plays a crucial role in the overall sustainability of construction projects. This subchapter explores the key principles and practices of sustainable geotechnical design and construction, providing engineers in the field of civil engineering with valuable insights and strategies.

One of the fundamental aspects of sustainable geotechnical design is the careful selection of materials. Engineers must consider the environmental impact of the materials used in construction, such as the extraction process, energy consumption, and carbon emissions. The use of recycled or locally sourced materials can significantly reduce the environmental footprint. Additionally, engineers should focus on minimizing waste generation and promoting the reuse or recycling of materials whenever possible.

Another crucial aspect is the management of construction sites. Implementing effective erosion and sediment control measures can prevent soil erosion and the discharge of pollutants into nearby water bodies. Proper site grading and drainage design can also reduce the risk of soil erosion and improve the stability of slopes. Additionally, engineers should aim to minimize the disturbance to existing ecosystems and habitats during construction, preserving the natural environment.

In terms of design, engineers should strive for innovative and sustainable solutions. Incorporating geotechnical techniques such as soil stabilization, ground improvement, and geosynthetics can enhance the performance and longevity of structures, reducing the need for frequent maintenance and repairs. Furthermore, engineers should consider the potential impacts of climate change on geotechnical systems, such as increased rainfall or rising sea levels, and design structures accordingly to ensure long-term resilience.

Collaboration and communication among multidisciplinary teams are vital for sustainable geotechnical design and construction. Engineers should work closely with architects, environmental experts, and other stakeholders to integrate sustainable practices into the overall project design. This collaborative approach allows for the identification of potential environmental risks and the development of mitigation strategies, ensuring the successful implementation of sustainable geotechnical solutions.

In conclusion, sustainable geotechnical design and construction practices are essential for engineers in the field of civil engineering. By incorporating environmentally friendly materials, managing construction sites responsibly, and implementing innovative design solutions, engineers can contribute to the long-term sustainability of construction projects. Through collaboration and communication, engineers can address the unique challenges of each project and create a more sustainable built environment.

Future Trends in Sustainable Geotechnical Engineering

As the world continues to grapple with the challenges of climate change and environmental degradation, the field of geotechnical engineering is evolving to meet the demands of sustainable development. In this subchapter, we will explore the future trends in sustainable geotechnical engineering, focusing on how these advancements can positively impact the field of civil engineering.

One of the key trends in sustainable geotechnical engineering is the integration of renewable energy technologies into geotechnical systems. Engineers are increasingly exploring the use of geothermal energy, solar panels, and wind turbines in geotechnical projects. These technologies not only provide clean energy but also have the potential to reduce the environmental impact of construction activities. By harnessing renewable energy, geotechnical engineers can contribute to a more sustainable and resilient built environment.

Another future trend is the adoption of advanced materials and techniques in geotechnical engineering. Innovations such as geosynthetics, biotechnical engineering, and self-healing materials are revolutionizing the field. Geosynthetics, for example, offer improved soil stabilization, erosion control, and drainage solutions. Biotechnical engineering involves using living organisms or their byproducts to enhance soil properties and strengthen slopes. Self-healing materials have the ability to repair cracks and damages, increasing the lifespan of geotechnical structures. These advancements not only improve the performance of geotechnical systems but also contribute to sustainable construction practices.

Furthermore, the concept of circular economy is gaining traction in geotechnical engineering. Instead of the traditional linear model of extracting resources, using them, and disposing of them, the circular economy aims to minimize waste and maximize resource efficiency. Geotechnical engineers can play a vital role in this transition by promoting the reuse and recycling of materials, implementing sustainable construction practices, and designing structures that can be repurposed or deconstructed at the end of their lifespan. By embracing the principles of the circular economy, geotechnical engineering can contribute to reducing carbon emissions and minimizing the depletion of natural resources.

Lastly, the future of sustainable geotechnical engineering lies in the integration of digital technologies. Building information modeling (BIM), remote sensing, and artificial intelligence (AI) are being increasingly used to optimize geotechnical design, construction, and monitoring processes. BIM allows engineers to create digital representations of geotechnical projects, facilitating collaboration and improving decision-making. Remote sensing techniques, such as LiDAR and satellite imagery, provide valuable data for site characterization and monitoring. AI algorithms can analyze large datasets, predict soil behavior, and optimize geotechnical designs. By embracing digital technologies, geotechnical engineers can enhance the sustainability and efficiency of their projects.

In conclusion, the future of sustainable geotechnical engineering holds great promise for the field of civil engineering. The integration of renewable energy technologies, advanced materials, circular economy principles, and digital technologies will enable engineers to design and construct more sustainable and resilient geotechnical systems. By staying abreast of these future trends, civil engineers can contribute to the sustainable development of our built environment and mitigate the impacts of climate change.

Chapter 12: Case Studies in Geotechnical Engineering

Case Study 1: Foundation Design for High-Rise Building

Introduction:
In the field of geotechnical engineering, the foundation design for high-rise buildings holds paramount importance. The stability and safety of these towering structures heavily rely on the proper selection and design of foundations. This case study explores the challenges and solutions encountered during the foundation design for a high-rise building, showcasing the principles and practices that engineers in civil engineering must employ.

Project   Background:
The case study focuses on a prestigious high-rise building project located in a major urban center. The construction site is characterized by complex soil conditions, including soft clay and loose sands. The engineers faced the challenge of designing a foundation system that could accommodate the building's weight, mitigate settlement, and withstand potential external forces such as earthquakes and wind loads.

Geotechnical   Investigation:
Prior to foundation design, a comprehensive geotechnical investigation was conducted. It involved soil sampling, laboratory testing, and in-situ testing to determine the soil properties and their behavior under load. The investigation revealed the presence of weak soil layers and potential liquefaction zones, emphasizing the need for a robust foundation design.

Foundation   System   Selection:
Based on the geotechnical investigation, various foundation

systems were considered, including shallow foundations, deep foundations, and innovative solutions. The engineers evaluated the pros and cons of each system, considering factors such as cost, time, and construction feasibility. Ultimately, a deep foundation system was selected to ensure the building's stability and minimize settlement issues.

Design Analysis and Calculation:
The design process involved extensive analysis and calculations to determine the appropriate dimensions and reinforcement requirements for the foundation elements. The engineers utilized advanced geotechnical software to model the complex soil-structure interaction and assess the building's response to various loads and ground conditions. The analysis considered factors such as bearing capacity, settlement, lateral stability, and seismic performance.

Construction Challenges and Solutions:
During construction, the engineers encountered challenges related to excavation, ground improvement, and the installation of foundation elements. They implemented innovative techniques such as soil stabilization using jet grouting, ground improvement through vibro-compaction, and the utilization of high-capacity piles to overcome these challenges and ensure the successful implementation of the foundation system.

Conclusion:
The foundation design for high-rise buildings demands a comprehensive understanding of geotechnical principles and practices. This case study highlights the critical role of geotechnical engineering in ensuring the stability and safety of such structures. By addressing complex soil conditions and employing innovative techniques, engineers in civil engineering can design and construct robust foundation systems that withstand the test of time.

Case Study 2: Stability Analysis of Slopes in Highway Construction

Introduction:
In the field of civil engineering, the stability of slopes in highway construction plays a crucial role in ensuring the safety and durability of transportation infrastructure. Slope failures can lead to costly damages, disruptions in traffic flow, and even loss of lives. Therefore, it is imperative for engineers to conduct thorough stability analyses to identify potential risks and implement appropriate mitigation measures. This case study presents a comprehensive analysis of slope stability in the context of highway construction, providing engineers with valuable insights and practical solutions.

Background:
Highway construction involves various geotechnical challenges, among which slope stability is a critical concern. The natural terrain often requires cutting into hillsides or building embankments, creating slopes that must withstand the forces of gravity, rainfall, and other external factors. Understanding the behavior of slopes and their potential failure mechanisms is essential for engineers to design safe and cost-effective highways.

Methods   and   Analysis:
This subchapter delves into the fundamental principles and practices of stability analysis for slopes in highway construction. It begins by discussing the importance of site investigations, which include geotechnical exploration and laboratory testing to determine soil properties and assess slope conditions. The chapter then introduces different methods of stability analysis, such as limit equilibrium methods and numerical modeling techniques, providing step-by-step guidance on their application.

Case   Study:
To illustrate the practical application of stability analysis, a comprehensive case study is presented. The case study examines a real-life highway construction project where slope stability issues arose. It explores the factors that contributed to the instability and the approach taken to mitigate the risks. Detailed analysis, including slope geometry, soil properties, and external forces acting on the slope, is conducted to determine the critical failure surfaces and assess the factor of safety.

Mitigation   Strategies:
Based on the analysis results, a range of mitigation strategies is proposed and evaluated. These may include slope reinforcement techniques, such as soil nailing or the installation of rock bolts, as well as slope drainage systems to control groundwater. The case study highlights the importance of selecting the most suitable mitigation measures based on site-specific conditions and project constraints.

Conclusion:
Stability analysis of slopes in highway construction is an essential aspect of geotechnical engineering. This subchapter provides engineers with a comprehensive understanding of stability analysis methods and their practical application through a real-life case study. By following the principles and practices outlined in this subchapter, engineers can effectively identify and mitigate slope stability risks, thus ensuring the long-term safety and durability of highway infrastructure.

Case Study 3: Retaining Wall Design for Waterfront Development

Introduction:

In this case study, we will explore the design and construction challenges faced in the development of a waterfront project that required the implementation of a retaining wall. The project was undertaken by a team of civil engineers who aimed to ensure the structural integrity and stability of the waterfront area while maximizing the available space for development.

Background:

Waterfront developments are becoming increasingly popular as cities strive to utilize their waterfront areas for commercial, residential, and recreational purposes. However, constructing near water bodies presents unique geotechnical challenges, such as soil erosion, water pressure, and potential slope instability. To address these challenges, the team was entrusted with designing and constructing a retaining wall that would provide the necessary stability and support for the project.

Design Considerations:

The engineers faced several design considerations while planning the retaining wall. Firstly, they needed to assess the soil conditions and evaluate the potential for erosion, which is common in waterfront areas. To counteract this, proper drainage systems were incorporated into the design to prevent water buildup behind the wall.

Secondly, the engineers had to determine the appropriate type of retaining wall based on the site conditions, load requirements, and aesthetic considerations. Various types, including gravity walls, cantilever walls, and sheet pile walls, were evaluated. Ultimately, the team opted for a cantilever retaining wall due to its cost-effectiveness, stability, and ease of construction.

Construction Challenges:

During the construction phase, the engineers encountered several challenges. The site's limited access, proximity to water, and high water table complicated the construction process. Specialized equipment and techniques were employed to overcome these obstacles, ensuring the safety of the workers and the structural integrity of the retaining wall.

Conclusion:

The successful design and construction of the retaining wall for this waterfront development project exemplify the importance of geotechnical engineering in civil engineering practices. Engineers must consider various factors, such as soil conditions, load requirements, and construction challenges, to develop effective and sustainable solutions.

By addressing these challenges, the team ensured the stability and longevity of the retaining wall while providing a solid foundation for the waterfront development. This case study serves as a valuable reference for civil engineers involved in similar projects, highlighting the significance of geotechnical engineering principles and practices in waterfront construction.

Case Study 4: Ground Improvement Techniques in Soft Soil

Soft soil poses significant challenges in construction projects, particularly in the field of civil engineering. This case study explores various ground improvement techniques that engineers can employ to address the specific issues associated with soft soil.

Soft soil is characterized by low bearing capacity, high compressibility, and poor drainage properties. These factors can lead to settlement, instability, and even failure of structures built on such soil. To overcome these challenges, engineers must implement appropriate ground improvement techniques to enhance the properties of the soil.

One effective method of ground improvement is the use of deep soil mixing. This technique involves the mechanical blending of cementitious materials into the soil, creating a stronger and more stable foundation. Deep soil mixing is particularly suitable for soft soil as it improves both the shear strength and compressibility of the ground.

Another commonly employed technique is the installation of stone columns or piers. These columns are constructed by inserting compacted stone aggregates into the soft soil, creating a reinforced ground structure. Stone columns not only increase the bearing capacity of the soil but also improve its drainage characteristics by creating a vertical drainage path.

Additionally, soil stabilization using chemical additives can be employed to improve the properties of soft soil. This technique involves the injection of chemical agents into the ground to enhance its strength and stability. Lime and cement are commonly used additives that chemically react with the soil, increasing its shear strength and reducing its compressibility.

Furthermore, engineers can implement ground improvement through the use of geosynthetics. Geotextiles and geogrids are placed within the soil to provide reinforcement and improve its overall stability. Geosynthetics can also be used for soil separation, filtration, and drainage, thus enhancing the performance of soft soil in construction projects.

In conclusion, soft soil presents significant challenges in civil engineering projects. However, by employing appropriate ground improvement techniques such as deep soil mixing, stone columns, soil stabilization, and geosynthetics, engineers can overcome these challenges and ensure the successful completion of construction projects. It is essential for engineers to have a comprehensive understanding of these techniques to effectively address the specific issues associated with soft soil and ensure the longevity and stability of structures.

Case Study 5: Seismic Design of Bridges and Infrastructure

Introduction:

In the field of civil engineering, the seismic design of bridges and infrastructure plays a crucial role in ensuring the safety and reliability of these structures during earthquakes. The dynamic nature of seismic forces requires engineers to adopt specialized design techniques and considerations to mitigate potential damage and ensure the durability of these structures. This case study delves into the seismic design principles and practices employed in the construction of bridges and other infrastructure, shedding light on the challenges faced by engineers and the innovative solutions they have developed.

Seismic Design Principles:

Seismic design begins with a thorough understanding of the local geotechnical conditions and seismic hazard analysis. Engineers must assess the site-specific ground motion parameters, such as peak ground acceleration and spectral response, to determine the design forces that the structures will be subjected to during an earthquake. This data is crucial in determining the appropriate design criteria, including the selection of suitable soil-structure interaction models and seismic design codes.

Design Considerations:

The seismic design of bridges and infrastructure involves several critical considerations. Engineers must focus on designing structures with sufficient strength, stiffness, and ductility to resist the lateral forces induced by seismic events. Special attention is given to the selection and placement of reinforcement, with an emphasis on providing adequate anchorage and confinement to prevent structural failure. Additionally, engineers must consider the potential for soil liquefaction, landslides, and other geotechnical hazards that can significantly affect the performance of these structures during earthquakes.

Innovative Solutions:

Over the years, engineers have developed innovative solutions to enhance the seismic resistance of bridges and infrastructure. These include the use of base isolation systems, which decouple the structure from the ground, dissipating seismic energy and reducing the impact of ground motion on the structure. Other advancements include the development of energy dissipation devices, such as dampers and viscoelastic materials, which absorb and dissipate seismic energy, protecting the structural integrity.

Conclusion:

The seismic design of bridges and infrastructure is a complex and challenging task that requires the expertise of civil engineers. By incorporating site-specific seismic hazard analysis, appropriate design criteria, and innovative solutions, engineers can ensure the safety and resilience of these structures during earthquakes. The lessons learned from case studies like this help advance the field of geotechnical engineering, enabling engineers to design structures that can withstand the forces of nature and protect lives and critical infrastructure.

Chapter 13: Emerging Technologies in Geotechnical Engineering

Introduction to Emerging Technologies in Geotechnical Engineering

In recent years, the field of geotechnical engineering has witnessed significant advancements due to emerging technologies. These technologies have not only enhanced the accuracy and efficiency of geotechnical investigations and analyses but have also revolutionized the way engineers approach various geotechnical challenges in construction projects. This subchapter aims to introduce engineers, specifically those in the niche of civil engineering, to some of the most promising emerging technologies in geotechnical engineering.

One of the key emerging technologies in geotechnical engineering is remote sensing. Remote sensing techniques, such as LiDAR (Light Detection and Ranging) and aerial imaging, provide engineers with high-resolution data about the land surface and subsurface conditions. By analyzing these data, engineers can accurately assess potential geotechnical hazards, identify suitable locations for construction, and design appropriate foundation systems.

Another significant advancement is the application of unmanned aerial vehicles (UAVs) or drones in geotechnical engineering. UAVs equipped with various sensors and cameras can quickly and safely collect data from difficult-to-access areas, such as steep slopes or hazardous terrains. This data can be used to create detailed 3D models, monitor slope stability, and detect any deformations or displacements in real-time.

Geotechnical instrumentation has also progressed with the introduction of wireless sensor networks (WSNs). WSNs enable real-time monitoring of geotechnical parameters, such as pore water pressure, settlement, and slope stability. These sensors can be embedded in the ground or attached to structures, providing engineers with continuous and accurate data. This real-time monitoring allows for early detection of potential geotechnical problems, ensuring timely and appropriate interventions.

The use of advanced geotechnical modeling software is another emerging technology that has significantly improved the design and analysis of geotechnical structures. These software tools utilize numerical methods, such as finite element or finite difference methods, to simulate complex geotechnical processes accurately. Engineers can analyze the behavior of soil, rock, and structures under various loading conditions, leading to more efficient and cost-effective designs.

Furthermore, the integration of artificial intelligence (AI) and machine learning (ML) algorithms in geotechnical engineering has shown great potential. AI and ML techniques can analyze vast amounts of geotechnical data, identify patterns, and make predictions. This enables engineers to optimize designs, predict geotechnical hazards, and make informed decisions based on historical data.

In conclusion, emerging technologies in geotechnical engineering have opened up new possibilities for engineers in the civil engineering niche. Remote sensing, UAVs, wireless sensor networks, advanced geotechnical modeling software, and AI/ML algorithms have all contributed to improving the accuracy, efficiency, and safety of geotechnical investigations and designs. It is crucial for engineers to stay updated with these emerging technologies to enhance their capabilities and deliver better geotechnical solutions in construction projects.

Remote Sensing and Geotechnical Applications

In recent years, remote sensing technology has revolutionized the field of geotechnical engineering, providing engineers with invaluable data and insights for construction projects. This subchapter explores the various applications of remote sensing in geotechnical engineering and its significance in the field of civil engineering.

Remote sensing involves the use of satellite, aerial, and ground-based sensors to collect data about the Earth's surface without direct contact. This technology allows engineers to obtain critical information about the subsurface conditions, topography, and environmental factors that can significantly impact construction projects.

One of the key applications of remote sensing in geotechnical engineering is its ability to gather data on ground conditions prior to construction. By analyzing satellite images and aerial photographs, engineers can identify geological formations, soil types, and potential hazards such as landslides or sinkholes. This information is crucial for accurate planning and design, ensuring the safety and stability of the structures to be built.

Remote sensing also plays a vital role in monitoring construction sites. Engineers can use satellite imagery and ground-based sensors to track ground movements, settlement, and deformation during different phases of the construction process. This real-time data allows for early detection of potential issues, enabling engineers to take necessary measures to prevent accidents or structural failures.

Moreover, remote sensing techniques assist in environmental monitoring and assessment. By analyzing satellite images and thermal data, engineers can evaluate the impact of construction activities on the surrounding ecosystem, including changes in vegetation, water bodies, and air quality. This information is invaluable for sustainable construction practices and complying with environmental regulations.

Additionally, remote sensing technology enables engineers to assess the long-term performance of geotechnical structures. By continuously monitoring settlements, slope stability, and soil moisture content through remote sensing, engineers can identify any signs of deterioration or potential failures. This proactive approach allows for timely maintenance and repair, ensuring the safety and longevity of infrastructure.

In conclusion, remote sensing has emerged as an indispensable tool in geotechnical engineering, providing engineers with critical data for planning, design, monitoring, and maintenance of construction projects. Its applications span from site characterization to environmental assessment, making it an essential component of civil engineering practices. By leveraging remote sensing technology, engineers can enhance the safety, efficiency, and sustainability of geotechnical projects, ultimately contributing to the advancement of the field as a whole.

Artificial Intelligence and Machine Learning in Geotechnical Engineering

Artificial Intelligence (AI) and Machine Learning (ML) have revolutionized various fields, and geotechnical engineering is no exception. In recent years, AI and ML have emerged as powerful tools for solving complex problems and making informed decisions in civil engineering, particularly in the geotechnical domain. This subchapter explores the potential applications and benefits of AI and ML in geotechnical engineering, providing engineers in the civil engineering niche with a comprehensive understanding of these technologies.

AI and ML techniques can enhance the efficiency and accuracy of traditional geotechnical engineering practices. For instance, these technologies can analyze large volumes of geotechnical data, including soil properties, site conditions, and historical records, to identify patterns and trends that might not be apparent to human engineers. By leveraging AI and ML algorithms, engineers can gain valuable insights into soil behavior, groundwater movement, and other critical factors that influence the stability and performance of construction projects.

One of the significant advantages of AI and ML is their ability to predict and mitigate geotechnical hazards. These technologies can process real-time data from sensors embedded in the ground or collected through remote monitoring systems. By continuously analyzing this data, AI and ML algorithms can detect early warning signs of potential failures, such as slope instability or soil liquefaction. This predictive capability enables engineers to implement timely preventive measures, reducing the risk of accidents and minimizing project delays.

Furthermore, AI and ML algorithms can optimize the design and construction process of geotechnical projects. By considering numerous design parameters, such as soil types, site conditions, and load requirements, these algorithms can generate optimized solutions that meet project constraints while minimizing costs and environmental impact. This automated design process not only saves time but also ensures that geotechnical structures are designed with maximum efficiency and safety.

However, it is crucial for engineers to understand the limitations and potential pitfalls of AI and ML in geotechnical engineering. While these technologies offer tremendous potential, they rely heavily on the quality and representativeness of the input data. Engineers must exercise caution and validate the results obtained from AI and ML algorithms with conventional geotechnical engineering methods.

In conclusion, AI and ML have the potential to revolutionize geotechnical engineering practices in the civil engineering niche. By leveraging these technologies, engineers can enhance the accuracy of soil analysis, predict geotechnical hazards, and optimize the design and construction of geotechnical projects. However, engineers must remain vigilant and supplement AI and ML techniques with traditional geotechnical engineering methods to ensure reliable and safe project outcomes.

Robotics and Automation in Geotechnical Construction

In recent years, the field of geotechnical engineering has witnessed a significant transformation with the integration of robotics and automation. This subchapter explores the growing significance of robotics and automation in geotechnical construction, addressing the audience of engineers, particularly those specializing in civil engineering.

The advent of robotics and automation has revolutionized the way geotechnical construction projects are executed. These technologies offer numerous advantages, including increased productivity, improved safety, and enhanced precision. With the ability to perform tasks with high accuracy and repeatability, robots and automated systems have become indispensable tools in geotechnical engineering.

One of the key areas where robotics and automation have revolutionized geotechnical construction is in site investigation and data collection. Robotic devices equipped with sensors and cameras can collect data from remote and hazardous locations, eliminating the need for human intervention in potentially dangerous environments. This allows for more comprehensive and accurate data collection, leading to better-informed decision-making during the design and construction phases.

Furthermore, robots and automated systems have greatly enhanced the efficiency and effectiveness of construction techniques such as drilling, pile driving, and excavation. Automated drilling rigs can precisely control drilling parameters, leading to improved hole quality and reduced project timelines. Similarly, automated pile drivers can accurately drive piles into the ground, ensuring optimal load-bearing capacity and minimizing the risk of structural failures.

Automation has also significantly impacted the field of slope stability analysis and monitoring. Robotic devices equipped with advanced sensors can collect real-time data on slope movements, allowing engineers to detect potential instabilities and take timely preventive measures. This proactive approach to slope stability analysis can prevent catastrophic failures and ensure the long-term safety of geotechnical structures.

Moreover, robotics and automation have opened new possibilities in geotechnical construction by enabling remote and autonomous operations. Unmanned aerial vehicles (UAVs) equipped with cameras and LiDAR scanners can capture high-resolution images and topographic data, facilitating accurate site mapping and monitoring. Autonomous vehicles can transport materials and equipment on construction sites, reducing human labor and enhancing overall project efficiency.

In conclusion, the integration of robotics and automation in geotechnical construction has revolutionized the industry, offering engineers in civil engineering a wide range of benefits. From improved data collection and analysis to enhanced construction techniques and remote operations, these technologies have reshaped the field of geotechnical engineering. Embracing robotics and automation is crucial for engineers to stay ahead in the modern construction industry and ensure the successful and safe execution of geotechnical projects.

3D Printing and Additive Manufacturing in Geotechnical Engineering

The field of geotechnical engineering has witnessed significant advancements in recent years, thanks to the emergence of 3D printing and additive manufacturing technologies. These groundbreaking technologies have revolutionized the way civil engineers approach various challenges in construction projects, offering new possibilities and solutions that were previously unimaginable.

One of the key advantages of 3D printing and additive manufacturing in geotechnical engineering lies in their ability to create complex geometries accurately and efficiently. Traditional construction methods often face limitations when it comes to shaping intricate designs or dealing with irregular soil conditions. However, with 3D printing, engineers can now produce custom-designed structures, such as retaining walls, tunnels, and even foundations, with ease.

Moreover, additive manufacturing techniques allow for the use of a wide range of materials, including advanced composites and soil-based mixtures, tailored to specific geotechnical requirements. This flexibility enhances the performance and durability of geotechnical structures, ensuring their stability and resilience in challenging environmental conditions.

Another significant advantage of 3D printing and additive manufacturing in geotechnical engineering is the potential to reduce construction time and costs. Traditional construction methods often involve extensive excavation, formwork, and manual labor, which can be time-consuming and expensive. By contrast, 3D printing enables the rapid fabrication of geotechnical structures, minimizing the need for extensive site preparation and reducing overall project duration. This not only saves time but also reduces labor costs, making construction projects more economical.

Furthermore, the use of 3D printing in geotechnical engineering offers improved sustainability and environmental benefits. Additive manufacturing techniques generate less waste compared to traditional construction methods, as they only use the necessary amount of material. Additionally, 3D printing allows for the use of recycled materials, contributing to a more environmentally friendly approach to construction.

In conclusion, 3D printing and additive manufacturing have ushered in a new era in geotechnical engineering, providing engineers in the civil engineering niche with unparalleled opportunities. By leveraging these technologies, engineers can design and construct geotechnical structures with complex geometries, tailored materials, and reduced construction time and costs. As the field continues to evolve, it is imperative for engineers to embrace and explore the potentials of 3D printing and additive manufacturing to drive innovation and shape the future of geotechnical engineering in construction.

Chapter 14: Professional Ethics and Responsibilities in Geotechnical Engineering

Importance of Professional Ethics in Engineering

In the field of civil engineering, professional ethics play a crucial role in maintaining the integrity, reputation, and safety of the profession. Engineers have a responsibility to uphold ethical standards in their work, as their decisions and actions can have significant impacts on society and the environment. This subchapter aims to highlight the importance of professional ethics in engineering, focusing on the niche of civil engineering.

First and foremost, professional ethics ensure the safety and well-being of the public. Civil engineers design and construct infrastructure that directly affects the lives of people. From bridges and roads to buildings and dams, the structures built by civil engineers must adhere to strict safety standards. By adhering to professional ethics, engineers prioritize public safety over personal gain, ensuring that their work is of the highest quality and meets all relevant regulations.

Professional ethics also help maintain the reputation of the civil engineering profession. Clients, stakeholders, and the general public place their trust in engineers and expect them to act with integrity. By following ethical guidelines, engineers demonstrate their commitment to honesty, transparency, and accountability. This, in turn, enhances the reputation of the profession and fosters trust among stakeholders, leading to increased opportunities and collaborations.

Additionally, professional ethics promote sustainable and environmentally conscious practices in civil engineering.

Engineers must consider the long-term impacts of their projects on the environment and strive to minimize any negative effects. Ethical considerations such as reducing waste, conserving resources, and implementing sustainable design principles are essential in ensuring a greener and more sustainable future.

Moreover, professional ethics in engineering encourage continuous professional development. Engineers are responsible for keeping up with the latest advancements in technology, safety regulations, and industry best practices. By adhering to ethical standards, engineers are motivated to pursue ongoing education and training, ensuring that their knowledge and skills remain up-to-date.

In conclusion, professional ethics are of utmost importance in the field of civil engineering. They ensure the safety of the public, maintain the reputation of the profession, promote sustainable practices, and encourage continuous professional development. Civil engineers must adhere to ethical guidelines to ensure that their work is of the highest quality, upholds integrity, and contributes to the betterment of society. By doing so, engineers can make a positive impact on the world and leave a lasting legacy for future generations.

Code of Ethics for Geotechnical Engineers

As geotechnical engineers, it is essential for us to adhere to a strong code of ethics that guides our professional conduct and ensures the safety, integrity, and sustainability of civil engineering projects. The following code of ethics outlines the principles and practices that every geotechnical engineer should abide by:

1. Professional Integrity: Geotechnical engineers must maintain the highest standards of professional integrity, honesty, and impartiality in all their dealings. They should prioritize the public interest, safety, and welfare above all personal or financial considerations.

2. Competence: Geotechnical engineers should only undertake projects or provide services for which they are qualified by education, training, and experience. They must continuously improve their knowledge and skills to stay updated with the latest advancements in the field of geotechnical engineering.

3. Safety and Risk Management: Geotechnical engineers must prioritize the safety of the public and workers involved in construction projects. They should assess and manage risks associated with geotechnical hazards such as landslides, soil liquefaction, and foundation failures. Additionally, they should communicate potential risks effectively to all stakeholders.

4. Environmental Stewardship: Geotechnical engineers should promote sustainable practices and minimize adverse environmental impacts during construction projects. They should consider the long-term effects of their designs on the natural environment and strive to protect and preserve natural resources.

5. Professional Development and Mentoring: Geotechnical engineers should actively participate in professional societies and organizations to promote the exchange of knowledge and expertise. They should also mentor and guide young engineers to ensure the continuity of ethical practices and the development of future geotechnical engineering professionals.

6. Confidentiality: Geotechnical engineers must maintain the confidentiality of client information and proprietary data. They should not disclose any confidential information without proper authorization, ensuring the privacy and trust of their clients.

7. Ethical Communication and Reporting: Geotechnical engineers should provide accurate and truthful information in their reports, presentations, and communications. They should avoid misrepresentation or exaggeration of data and findings, promoting transparency and accountability in their work.

By following this code of ethics, geotechnical engineers contribute to the advancement of civil engineering and ensure the highest level of professionalism and integrity in their practice. Upholding these principles not only benefits the engineers themselves but also the communities they serve and the environment they impact.

Professional Responsibilities in Geotechnical Engineering

In the field of geotechnical engineering, professionals play a crucial role in ensuring the safety, stability, and sustainability of construction projects. Geotechnical engineers are responsible for assessing the physical properties of the soil and rock that will support these structures. They provide valuable insights into the behavior of the ground, helping to mitigate risks and design foundations that can withstand the forces imposed on them.

As engineers in the civil engineering niche, it is essential to understand the professional responsibilities that come with working in geotechnical engineering. These responsibilities encompass various aspects, including ethical considerations, project planning and management, and effective communication.

Ethical responsibilities are paramount in geotechnical engineering. Engineers must adhere to a strict code of ethics that promotes honesty, integrity, and the protection of public welfare. They must prioritize safety over any other considerations and ensure that their designs and recommendations are based on sound engineering principles and thorough analysis.

Additionally, geotechnical engineers have a responsibility to diligently plan and manage projects. This includes conducting thorough site investigations to gather essential data about the subsurface conditions, analyzing the data to assess potential risks and challenges, and designing appropriate foundations or ground improvement techniques. They must also consider the long-term sustainability of the project, taking into account environmental impacts and ensuring the proper disposal of excavated materials.

Effective communication is another vital responsibility for geotechnical engineers. They must be able to clearly convey their findings, recommendations, and limitations to other professionals, stakeholders, and clients who may not have a technical background. Clear communication is crucial to ensure that all parties involved in the project have a comprehensive understanding of the geotechnical aspects and can make informed decisions.

In conclusion, professional responsibilities in geotechnical engineering are significant and require a meticulous approach. As civil engineers, it is essential to uphold ethical standards, diligently plan and manage projects, and communicate effectively with all stakeholders. By fulfilling these responsibilities, geotechnical engineers contribute to the safe and successful construction of infrastructure projects, ensuring the stability and longevity of our built environment.

Ethical Decision Making in Geotechnical Engineering

Introduction:
In the field of geotechnical engineering, ethical decision making plays a crucial role in ensuring the safety, sustainability, and success of construction projects. Civil engineers, with their expertise in designing and constructing infrastructure, have a responsibility to uphold ethical standards to protect public welfare. This subchapter explores the importance of ethical decision making in geotechnical engineering and provides guidelines for engineers to navigate the complex ethical dilemmas they may encounter in their profession.

1. Importance of Ethical Decision Making: Geotechnical engineering involves analyzing soil properties, designing foundations, and mitigating geotechnical risks. Engineers must make decisions that prioritize safety, environmental sustainability, and long-term stability. Ethical decision making ensures that engineers consider the potential consequences of their actions and make choices that align with professional standards and societal expectations.

2. Ethical Principles in Geotechnical Engineering: This section discusses the ethical principles that engineers should consider when making decisions in geotechnical engineering. These principles include honesty, integrity, competence, accountability, and professionalism. Engineers must prioritize transparency, ensure their qualifications match the task at hand, take responsibility for their actions, and maintain a high level of professionalism in their interactions with clients, colleagues, and the public.

3. Ethical Dilemmas in Geotechnical Engineering: Geotechnical engineers often face complex ethical dilemmas, such as balancing cost constraints with safety requirements, dealing with conflicting project objectives, or addressing potential risks to the environment. This section explores these dilemmas and provides case studies to illustrate how engineers can navigate these challenges while upholding ethical standards.

4. Ethical Decision-Making Framework: To assist engineers in making ethical decisions, this section presents a practical framework that can guide their thinking. The framework includes steps such as identifying the problem, gathering relevant information, analyzing potential solutions, considering the consequences, and making a well-reasoned decision. Engineers are encouraged to consult professional codes of ethics, seek advice from colleagues, and engage in continuous professional development to enhance their ethical decision-making skills.

Conclusion:
Ethical decision making is a fundamental aspect of geotechnical engineering that ensures the safety, sustainability, and ethical practice of the profession. Civil engineers must be aware of the ethical principles and dilemmas specific to geotechnical engineering and apply a systematic framework to make informed decisions. By upholding ethical standards, engineers can contribute to the advancement of the field and build a reputation of trust and integrity within the civil engineering community.

Case Studies on Ethical Issues in Geotechnical Engineering

Introduction:
Ethical issues in geotechnical engineering play a crucial role in the construction industry, particularly in civil engineering projects. Engineers need to be aware of the potential ethical dilemmas they may face in their profession and understand the importance of making ethical decisions to ensure the safety and well-being of society. This subchapter explores various case studies that shed light on ethical issues encountered by geotechnical engineers in their day-to-day work.

1. Case Study: Slope Stability Analysis:
In this case study, we examine an incident where a geotechnical engineer was faced with a dilemma while conducting slope stability analysis for a construction project. The engineer had discovered that the slope had a high risk of failure but was pressured by the project manager to manipulate the data to downplay the risks. The case study delves into the ethical implications of this situation and the potential consequences of compromising professional integrity.

2. Case Study: Foundation Design:
This case study focuses on a geotechnical engineer who faced an ethical dilemma while designing the foundation for a high-rise building. The engineer realized that the proposed design was inadequate to withstand the expected loads and might lead to catastrophic failure. However, the client, driven by financial constraints, insisted on moving forward with the inadequate design. The case study explores the engineer's ethical responsibilities and the potential legal implications of their decision.

3. Case Study: Environmental Impact Assessment: Geotechnical engineers often encounter ethical challenges when conducting environmental impact assessments. This case study examines a scenario where an engineer was asked to overlook potential environmental hazards associated with a construction project. The engineer had to weigh the economic benefits against the potential harm to the environment and public health, highlighting the importance of ethical decision-making in such situations.

Conclusion:
The case studies presented in this subchapter offer valuable insights into the ethical issues faced by geotechnical engineers in civil engineering projects. They emphasize the importance of upholding professional integrity and making ethical decisions to ensure the safety, sustainability, and well-being of the communities we serve. By understanding and learning from these case studies, engineers can enhance their ethical awareness and develop the skills necessary to navigate such challenging situations in their professional careers.

Chapter 15: Future Perspectives in Geotechnical Engineering

Current Challenges and Future Trends in Geotechnical Engineering

Introduction:
Geotechnical engineering plays a crucial role in the construction industry, particularly in civil engineering projects. This subchapter explores the current challenges faced by geotechnical engineers and provides insights into the future trends that will shape the field. As engineers in the civil engineering niche, it is essential to stay abreast of these challenges and trends to ensure safe and efficient construction practices.

1. Current Challenges:
a) Sustainability and Environmental Concerns: Geotechnical engineers are increasingly tasked with finding sustainable solutions that minimize environmental impact. This includes reducing carbon footprints, managing waste materials, and considering the long-term effects of construction on ecosystems.

b) Urbanization and Infrastructure Development: Rapid urbanization presents unique challenges for geotechnical engineers. The need for efficient infrastructure development often results in complex projects on challenging terrains, requiring innovative design and construction techniques.

c) Aging Infrastructure: Many countries are faced with aging infrastructure, including bridges, dams, and tunnels. Geotechnical engineers are responsible for assessing and managing the structural integrity of these aging assets, ensuring public safety.

d) Natural Hazards: Climate change-induced events such as floods, hurricanes, and earthquakes pose significant challenges to geotechnical engineers. Understanding the behavior of soils and mitigating the risks associated with these hazards are essential for sustainable infrastructure development.

2.   Future   Trends:
a) Advanced Computational Tools: The future of geotechnical engineering lies in the development and utilization of advanced computational tools. These tools, including machine learning algorithms and artificial intelligence, will enhance the analysis and design capabilities of engineers, leading to more accurate predictions and efficient designs.

b) Sustainable Geotechnical Solutions: As sustainability becomes a global priority, geotechnical engineers will play a crucial role in developing eco-friendly solutions. This includes utilizing recycled materials, adopting green construction practices, and implementing geothermal energy systems.

c) Geotechnical Monitoring Systems: Real-time monitoring systems will become increasingly important in geotechnical engineering. These systems will enable engineers to assess the performance of structures and monitor any potential geotechnical failures, allowing for timely interventions and risk mitigation.

d) Resilient Infrastructure: With the increasing frequency of natural disasters, geotechnical engineers will focus on designing resilient infrastructure that can withstand extreme events. This includes incorporating measures such as slope stability analysis, liquefaction mitigation, and innovative foundation systems.

Conclusion:
Geotechnical engineering in the civil engineering niche faces several current challenges and future trends. As engineers, it is crucial to address sustainability, urbanization, aging infrastructure, and natural hazards. Embracing advanced computational tools, sustainable solutions, geotechnical monitoring systems, and resilient infrastructure will shape the future of the field. By staying informed about these challenges and trends, engineers can contribute to the development of safe, efficient, and sustainable construction practices.

Advances in Geotechnical Engineering Research

Geotechnical engineering plays a fundamental role in the field of civil engineering, providing the necessary knowledge and tools to understand and design structures that interact with the ground. As technology continues to evolve, so does the research in geotechnical engineering, leading to significant advancements in this field. In this subchapter, we will explore some of the recent breakthroughs and their implications in construction practices, with a focus on addressing engineers in the niche of civil engineering.

One of the notable advances in geotechnical engineering research is the development of new and improved testing techniques. Traditional laboratory tests, such as triaxial testing and direct shear testing, have been enhanced with the inclusion of advanced instrumentation and data acquisition systems. These upgrades allow for more accurate measurements and analysis of soil behavior, enabling engineers to make better-informed decisions during the design and construction phases.

Another area of advancement is the application of remote sensing technologies in geotechnical engineering. Remote sensing techniques, such as LiDAR (Light Detection and Ranging) and satellite imagery, provide valuable information about the ground conditions over large areas. This data can be used to identify potential hazards, monitor slope stability, and assess the impact of natural disasters on infrastructure. By incorporating remote sensing technologies into geotechnical investigations, engineers can minimize risks and optimize construction processes.

Furthermore, the development of advanced numerical modeling software has revolutionized the way geotechnical engineers analyze and predict soil behavior. These software tools allow for complex simulations of ground response under various loading conditions, providing insights into the performance of geotechnical structures. By simulating real-world scenarios, engineers can optimize the design of foundations, retaining walls, and underground structures, ensuring their long-term stability and performance.

Additionally, geotechnical engineering research has witnessed advancements in the field of geosynthetics. Geosynthetics are synthetic materials used to improve the mechanical properties of soils, providing reinforcement, drainage, or filtration capabilities. Recent research has focused on the development of new materials and techniques to enhance the performance and durability of geosynthetics. These advancements have led to the design of more efficient and cost-effective geotechnical solutions, reducing construction time and minimizing environmental impacts.

In conclusion, the field of geotechnical engineering is constantly evolving, driven by research and technological advancements. Engineers in the niche of civil engineering can benefit greatly from staying updated with the latest developments in this field. The advances in testing techniques, remote sensing technologies, numerical modeling software, and geosynthetics discussed in this subchapter provide a glimpse into the exciting possibilities and opportunities that lie ahead in geotechnical engineering research. By incorporating these advances into their practices, engineers can improve the safety, efficiency, and sustainability of their construction projects.

Technological Innovations in Geotechnical Engineering

In recent years, the field of geotechnical engineering has witnessed significant advancements and innovations, revolutionizing the way engineers approach construction projects. These technological breakthroughs have not only enhanced the efficiency and accuracy of geotechnical investigations but also improved the overall safety and sustainability of civil engineering projects. This subchapter explores some of the most notable technological innovations in geotechnical engineering that are transforming the industry.

One of the most groundbreaking innovations in this field is the use of remote sensing technologies. Engineers can now utilize aerial photography, LiDAR (Light Detection and Ranging), and satellite imagery to obtain detailed information about the site conditions, topography, and vegetation cover. These data sources provide engineers with a comprehensive understanding of the site, enabling them to make informed decisions during the design and construction phases.

Furthermore, the advent of advanced geophysical techniques has revolutionized subsurface investigations. Engineers now have access to ground-penetrating radar (GPR), seismic refraction, and electrical resistivity methods, allowing them to map subsurface soil and rock layers with unparalleled accuracy. These techniques provide invaluable information about soil properties, groundwater levels, and potential hazards such as buried utilities or voids, significantly reducing the risk of unforeseen issues during construction.

Another significant technological innovation is the development of advanced instrumentation for monitoring geotechnical structures. Engineers can now use remote monitoring systems equipped with sensors to continuously measure and analyze parameters such as settlement, deformation, and pore pressure in real-time. This real-time data enables prompt identification of potential issues, allowing engineers to take proactive measures and mitigate risks effectively.

Furthermore, the emergence of geotechnical modeling software has transformed the way engineers analyze and design geotechnical structures. These software packages utilize advanced numerical methods, such as finite element analysis and limit equilibrium methods, to simulate complex soil-structure interactions. By providing accurate predictions of soil behavior and structural response, engineers can optimize designs, minimize uncertainties, and ensure the safety and stability of geotechnical structures.

In conclusion, the technological innovations in geotechnical engineering have revolutionized the industry, empowering engineers to make informed decisions, enhance safety, and improve the overall sustainability of civil engineering projects. The integration of remote sensing technologies, advanced geophysical techniques, real-time monitoring systems, and geotechnical modeling software has transformed the way engineers approach geotechnical investigations, design, and construction. As the field continues to evolve, engineers must embrace these innovations and leverage their potential to deliver better and more sustainable infrastructure solutions.

Education

Education is the foundation of any profession, and for engineers, it holds even greater significance. In the field of civil engineering, education plays a crucial role in shaping the skills and knowledge required to design and construct safe and efficient structures. This subchapter explores the various aspects of education that are essential for engineers in the niche of civil engineering.

One of the key elements of education in civil engineering is a strong theoretical foundation. Engineers must possess a deep understanding of mathematics, physics, and other scientific principles that underpin the design and analysis of structures. This theoretical knowledge equips engineers with the ability to solve complex problems and make informed decisions during the construction process.

In addition to theoretical knowledge, practical experience is vital for civil engineers. Education should include hands-on training in laboratory settings and construction sites, allowing engineers to apply their theoretical knowledge to real-world scenarios. This practical experience helps engineers develop a keen eye for detail, an understanding of construction materials, and the ability to manage and supervise construction projects effectively.

Continuing education is another vital aspect in the field of civil engineering. The industry is constantly evolving, with new technologies, materials, and design methods being introduced regularly. Engineers must stay updated with these advancements to remain competitive and deliver the best outcomes for their projects. By attending seminars, workshops, and conferences, engineers can expand their knowledge base and stay ahead of the curve.

Furthermore, education in civil engineering should also emphasize the importance of ethical and sustainable practices. Engineers have a responsibility to design structures that are safe for the public and have minimal impact on the environment. By incorporating ethical considerations and sustainable design principles into their education, engineers can contribute to the overall well-being of society.

Ultimately, education in civil engineering is a lifelong process. It starts with a strong theoretical foundation, is enhanced through practical experience, and continues through ongoing professional development. By investing in education, engineers can continually improve their skills, stay updated with industry advancements, and contribute to the growth and development of the field of civil engineering.